A CULTURAL HISTORY OF INSECTS

VOLUME 3

A Cultural History of Insects
General Editor: Gene Kritsky

Volume 1
A Cultural History of Insects in Antiquity
Edited by Gene Kritsky

Volume 2
A Cultural History of Insects in the Medieval Age
Edited by Gene Kritsky

Volume 3
A Cultural History of Insects in the Renaissance
Edited by Sherilyn G.F. Smith

Volume 4
A Cultural History of Insects in the Age of Enlightenment
Edited by Matthew S. Lehnert

Volume 5
A Cultural History of Insects in the Age of Industry
Edited by Carol Anelli and Susan Fisher

Volume 6
A Cultural History of Insects in the Modern Age
Edited by Robert K.D. Peterson

A CULTURAL HISTORY OF INSECTS

IN THE RENAISSANCE

VOLUME 3

Edited by Sherilyn G.F. Smith

BLOOMSBURY ACADEMIC
LONDON • NEW YORK • OXFORD • NEW DELHI • SYDNEY

BLOOMSBURY ACADEMIC
Bloomsbury Publishing Plc
50 Bedford Square, London, WC1B 3DP, UK
1385 Broadway, New York, NY 10018, USA
29 Earlsfort Terrace, Dublin 2, Ireland

BLOOMSBURY, BLOOMSBURY ACADEMIC and the Diana logo
are trademarks of Bloomsbury Publishing Plc

First published in Great Britain 2024

Series design by Raven Design
Cover image: Ms 134 Illuminated letter 'A' and side border of flowers, from
a Book of Hours, *c.*1500 © Victoria & Albert Museum, London, UK/Bridgeman Images

A catalogue record for this book is available from the British Library.

Library of Congress Cataloging-in-Publication Data

LCCN 2023029669

ISBN: HB: 978-1-3500-0307-1
Pack: 978-1-3500-0321-7
ePDF: 978-1-3502-6662-9
eBook: 978-1-3502-6663-6

Series: The Cultural Histories Series

Typeset by Integra Software Services Pvt. Ltd.
Printed and bound in Great Britain

CONTENTS

ILLUSTRATIONS

FIGURES

TABLE

CONTRIBUTORS

Evangelina Blust, PhD, Professor Emerita, Green Mountain College, Vermont, USA

Michael Blust, PhD, Professor Emeritus, Green Mountain College, Vermont, USA

Darryl Caterine, PhD, Professor of Religious Studies, Le Moyne College, USA

Lucinda Cole, PhD, Research Associate Professor of English and Affiliate Professor, Institute for Sustainability, Energy, and Environment at the University of Illinois, Urbana-Champaign, USA

Gene Kritsky, PhD, Dean, School of Behavioral and Natural Sciences, Mount St. Joseph University, USA

Peter C. Mancall, PhD, Andrew W. Mellon Professor of the Humanities at the University of Southern California, USA

Ann M. Martinez, PhD, Associate Professor of English, Kent State University at Stark, USA

Elda Miriam Aldasoro Maya, PhD, El Colegio de la Frontera Sur, Mexico, CONACYT Researcher

Cynthia M. Pyle, PhD, Co-Chair of the Columbia University Seminar in the Renaissance and recipient of fellowships at the American Academy in Rome, the Netherlands Institute for Advanced Study, and the Center for Advanced Study in the Visual Arts (National Gallery, Washington, USA)

Sherilyn G.F. Smith, PhD, Associate Professor of Biological and Environmental Sciences, Le Moyne College, USA

Michelle Ziegler, Graduate Student in Integrative Studies: Cultural Heritage & Resource Management, Southern Illinois University, Edwardsville, USA

SERIES PREFACE

This six-volume set, *The Cultural History of Insects*, is a unique examination of how insects have infested the human experience. The set is divided into six time periods: Antiquity, the Medieval Age, the Renaissance, the Age of Enlightenment, the Age of Industry, and the Modern Age. Within each time period, eight topics are addressed: the state of entomological knowledge during the period; insects and disease; insects and food; insect products; insects' role in mythology and religion; insects as symbols; the portrayal of insects in literature and language; and the representation of insects in art. This structure is meant to allow readers to tailor their reading to their specific interests: for example, if insects' influence in the history of art is of primary interest, the reader can select the art chapter from each volume; if a reader wants to understand the impact of insects on a particular period in human history, then reading the volume on that time period would be appropriate.

The coverage of the topics presented is not encyclopedic. Presented are histories and case studies describing how insects influenced human culture and survival. The geographic focus is generally the Western Hemisphere, but details are included of geographic intersections in which cross-cultural influence plays an important role in various topics and time periods.

Insects were here before humans. They have influenced our evolution, eaten our food, provided us with raw materials, changed our history, and been the muses for our creativity. These six volumes illustrate how insects have helped to shape our culture over the past five thousand years.

Gene Kritsky,
Mount St. Joseph University,
Cincinnati, OH

Introduction

SHERILYN G.F. SMITH

> To develop a complete mind: Study the art of science; study the science of art. Learn how to see. Realize that everything connects to everything else.
>
> — Leonardo da Vinci

The Renaissance—the term denotes a period of human history that has a narrow definition, being associated with European culture, but was also broad in its scope as trade as well as biological and cultural exchange became global. "Renaissance" is a descriptor of a period in Western civilization often interpreted as a "rebirth" of the arts and philosophy, a shift from a religious to a humanistic secular worldview in Europe.[1] Much more than a simple "awakening" from the Middle Ages, the period saw a renewed interest in ancient classical texts, questioning of religious authority and practices of the Church, a movement toward a secular humanism and intellectual inquiry, greater emphasis on the arts, the invention of the printing press and subsequent dissemination of ideas both old and new. The period also saw the developing complexity of global exploration and trade, as well as the invasion, conquest, and exploitation of entire continents by Europeans. Scientific inquiry grew out of the close observation of nature and the human body, comparing those observations with what had been written by the ancients, and the compilation of information and use of images to organize that growing body of knowledge.

The time frame itself associated with the Renaissance is similarly complex; it is defined for this volume as the fourteenth through the sixteenth centuries (1300–1600 CE) but within various regions and for different aspects of human culture, the time frame is much more fluid. Further, some scholars view the fourteenth century as the end of the High Middle Ages, while others see the sixteenth century as the beginning of the Early Modern Era. Nonetheless, however one defines the Renaissance, the period from 1300 to 1600 was a time of remarkable change. It is also important to remember that elsewhere in the world these centuries did not have the same significance as a time of "awakening." For many cultures even the time frame was measured differently—in the Muslim world, the Far East, Africa, and the Americas. In these cultures, terms like "fourteenth century" and "Renaissance" were meaningless. Even so, the changes that took place in Europe would reverberate across the globe as European contact, colonization, and trade extended to every continent.

Of particular significance was the mandate to Christian heads of state in Europe issued by Pope Nicholas V, in the form of a papal bull in 1452, now known as the "Doctrine of Discovery." Under this doctrine of the Church, Christian rulers were instructed to "capture, vanquish, and subdue the saracens, pagans, and other enemies of Christ," to "put them into perpetual slavery," and "to take all their possessions and property" (Davenport 1917: 20–6). The result was the invasion and conquest of non-Christian lands—Africa and the Americas—and the subjugation and enslavement of indigenous peoples under colonial rule. The African slave trade that developed in the late sixteenth century, as well as the opening of the New World to colonization and the exploitation of resources, created an exchange of people, other animals, plants, insects, and insect-borne disease organisms between continents. Referred to as the Columbian Exchange, this series of events would permanently alter cultures and ecologies around the world.

During the Renaissance, insects moved from a place of less prominence in human culture to a position much more central to the cultural changes that give the period its name. Insects were critical to the growing economy as producers of goods such as honey and silk. Insects also began to be recognized as threats to human enterprises, primarily causing agricultural damage via the depredations of caterpillars ("worms") and swarms of locusts. Insects were also known to sting and bite, although their role as vectors (transmitters) of human disease remained largely unknown. Through their roles in disease transmission, insects profoundly reshaped the cultures that fell victim to bubonic plague, malaria, and other diseases, even though the etiologic agents were unknown.

The importance of some insects to European society was well known as the Medieval Age transitioned into the Renaissance. Beekeeping, for example, was an activity of considerable importance, and beekeepers recorded detailed descriptions of the bees, their habits, and "best practices" of the time. Similarly, silkworms were a source of income for those engaged in silk production for the textile industry. "Best practices" were even improved upon during this time for both of these insects. Scale insects were used as a source of red pigments, which had been known since ancient times, and they too were raised for that purpose.

Apart from those few economically important insects, insects in general did not merit much attention at the beginning of the fourteenth century.

Insects gained recognition as objects of study as the intellectual climate of the fifteenth and sixteenth centuries fostered curiosity about the natural world. As the classical descriptions of Aristotle and Pliny the Elder were translated and made more accessible, all elements of nature began to be observed, described, cataloged, and compared to what had been described in those classical texts (see Beavis 1988). No longer were insects simply decorative and used to illuminate manuscripts; they were beginning to be regarded as worthy of study on their own. This growing intellectual interest in nature led to the emergence of a more scientific approach to inquiry, that of natural history, including detailed observations of plant and animal species and their interactions. Such careful observations of insects, along with what became a passion for collecting among these sixteenth-century natural historians, produced encyclopedias of insect knowledge and initiated entomology as a field of study. Increasingly detailed depictions of insects (often along with plants on which they were observed), both in works of art and as a means to illustrate the various encyclopedic works on insects, helped to shape this nascent field of inquiry.

Symbolic and/or religious meaning had been attached to some insects from antiquity: dung beetles symbolizing rebirth and butterflies representing the soul, for example. New associations emerged and old ones changed as the Church underwent the upheaval of the Reformation. These symbolic associations were often expressed visually, in religious and secular paintings and other works of art. Finally, the "discovery" of the New World in the late fifteenth century opened up what indeed *was* a new world of insects (and other creatures) to be described; moreover, insects were brought back to Europe either accidentally or intentionally. In addition, cross-cultural exchanges occurred with Indigenous peoples, for whom insects were an important part of their beliefs and practices.

Our understanding of these many roles of insects during this period comes from a variety of sources. Some are primary, such as the direct mention of insects in ecclesiastical trials, records of locust plagues, treatises on the culture of silkworms and honeybees, and literary works. Other sources are indirect, including tax, legal, and business records dealing with insect products and their manufacture and trade. Modern studies of Renaissance diseases have benefited from archaeological discoveries of mass graves (plague pits), as well as other artifacts containing human and/or insect remains. Advances in techniques such as analysis of DNA and immunological proteins in human remains provide a greater understanding of the incidence of insect-vectored diseases and their impact on local communities and wider regions.

Paintings, drawings, illustrations, and other visual sources are also important for understanding society's views of and the importance attached to insects. Visual representations of insects in illuminated manuscripts as well as in works of art, such as paintings and tapestries, are indicators of the artist's cultural perspective and that of society as a whole. Details of such depictions often show both the beauty and uniqueness of the insects themselves, as well as possible symbolic meanings attached to them. The way in which insects are placed in a painting relative to the other elements, as well as their size and number, can convey meaning, suggesting either importance or insignificance, for example. Imaginative construction of "pseudoinsects" and insect chimeras, sometimes very strange, demonstrate creativity of thought and expression by the artists and can take symbolism into new realms. In contrast, realistic and detailed drawings and descriptions of insects (as well as of plants and other living organisms) based on observations provide documentation that these insects existed. Such depictions of insects and other organisms from a more "biological" point of view constitute "photographs" of the natural world and also speak to the knowledge of the artist.

Any consideration of the part insects played during the Renaissance must be done in the context of the other elements of the period, and a chronological approach is useful. The changes in European society and culture associated with the Renaissance are generally acknowledged to have begun in southern Europe around the Mediterranean and spread north and west through Europe. Many historians describe Florence, Italy, under the rule of the Medici family, as the birthplace of the Renaissance. Spreading first to other Italian cities, a new focus on humanistic and intellectual endeavors such as art, music, literature, and science continued to spread more widely throughout Europe. Views of the self, society, the Church, nature, knowledge, and even time itself[2] underwent transformation during this period.

Europe at the beginning of the fourteenth century was primarily agricultural, with an increasing population that demanded more food and more land on which to grow that food. Deforestation for conversion to farmland was widespread. A change in the climate, beginning around 1290 CE and continuing into the fourteenth century brought cold, wet summers and harsh winter conditions to many parts of Europe, as well as periodic food shortages and famine in some areas (Cantor 2001). Ponting (2011) describes catastrophic conditions that occurred across Europe between 1315 and 1322 as a result of successive crop failures, during which over half of the sheep and oxen died, there was no grain or forage, and people in some places turned to cannibalism to survive. These harsh conditions, resulting from what is referred to as the first phase of the Little Ice Age, "set the stage" for serious socioeconomic changes as well as the spread of plague in the next few decades.[3] Events over which people had little or no control would permanently alter European society and their economies.

The fourteenth century saw the expansion of cities, both in size and influence, as rulers took over territory and consolidated power, particularly in the Italian city-states. Banking, trade, and craft industries grew and saw the benefit of technological innovations as well as shifts in the socioeconomic structure. New technologies were developed and old technologies advanced in many areas of endeavor, from the invention of silk-spinning machines to the printing press with movable type, to the clock and many other mechanical contrivances. Increased trade, both within Europe and also between European countries and Middle Eastern and Asian countries, improved economic conditions while at the same time bringing differing cultures into closer contact. Serfdom gave way to a working class, and craftsmen (and women) gained power through an expanding guild system. Cities and city-states continued to increase their economic and political power as they encouraged the growth of industries and pursued increased trade. This was evident in the rise to prominence of cities such as Milan, Florence, Paris, Brussels, Amsterdam, and London.

The period of the Renaissance also witnessed upheavals within the Roman Catholic Church, e.g., the Western Schism (1378–1417) during which there were two popes (and at one point, three) claiming authority. The Crusades came to an end in the late fifteenth century, and Martin Luther published his "Ninety-five Theses" in 1517, marking the beginning of the Reformation. That break with the Catholic Church reflected a shift in worldview that included not only the rejection of papal authority but the rejection of the concept of God manifest in nature. Tied to the subsequent conflicts over religious beliefs as well as to political and economic factors, there were numerous uprisings (e.g., the Peasants' Revolt of the late fourteenth century), geopolitical conflicts (e.g., the Hundred Years War), the Counter-Reformation, and conflicts between Catholic and Protestant monarchies. These conflicts had often life-changing consequences for the farmers, merchants, and craftspeople who worked with insects and their products; for example, artists and artisans who worked in silk took their knowledge and skills to new regions as they fled war or religious persecution. The distribution of financial resources and power also underwent changes as a new entrepreneurial class of powerful merchants arose, competing for wealth and privilege, and contributing to the conflicts of the time. Meanwhile, some aristocratic and even royal families found themselves in financial straits from lost revenues. The search for new sources of income had major effects that ranged from shifts in agricultural or manufactured products to expansion of trade.

Changes also occurred in trade networks related to geopolitical upheaval in eastern Europe and the Near East, most importantly the capture of Constantinople by the Ottoman Turks in 1453 and the subsequent obstacles to trade with China. The Silk Road, a major trade route (actually a group of several overland routes) to China, became more difficult and expensive for European merchants to utilize, eventually leading to efforts to find a shorter way to the Far East. Consequently, one of the most significant and world-changing events of the Renaissance occurred with the "discovery" of the New

World by Christopher Columbus in 1492. That discovery was followed by the conquest and colonization of the New World by Europeans in the late fifteenth through the sixteenth century. Exploration and colonization brought intercontinental movement of animals including insects, plants, pathogens, and other components of the natural world.

How did insects impact the daily lives of people during this period? Insects are ubiquitous and very small, so it is tempting to assume that they had little significance in daily life. However, a closer consideration of the conditions in which people lived and how those conditions changed throughout these three centuries in Europe suggest that, noticed or not, insects had a dramatic impact on life and culture. Mitchell (2015: 406) summarizes the widespread occurrence of ectoparasites including human fleas (*Pulex irritans*), bedbugs (*Cimex lectularius*), and head, body, and pubic lice (*Pediculus humanus capitus*, *P. humanus corporis*, and *Phthirius pubis*, respectively) based on extensive archaeological work across Europe. The plague pandemic (Black Death) that swept across the world in the fourteenth century, with epidemics recurring periodically for three centuries afterward, had a catastrophic impact on human society and influenced changing thought and belief systems, even though the role of insects went unrecognized. Plague mortality also led to more bargaining power for the surviving workers, whose services could command higher wages. As a result, society moved from feudalism to a peasant-based system; trade increased, a working class and guild systems developed, and urban centers grew in population, power, and influence. "Domesticated" insects such as honeybees, silkworms, and scale insects (a source of dyes and pigments) provided significant benefits by improving economies that fueled the patronage of many cultural and artistic endeavors. These insects also directly contributed to cultural expression, especially in the visual arts: paintings, drawings, and tapestries which used pigments and some of which used silk. Interactions with insects also became more complex as insects affected many different aspects of human culture in a great variety of new ways throughout the fifteenth and sixteenth centuries. Some of those interactions included the increased availability of many different silk textiles for clothing (fashion), new jobs related to production and the trade of insect products, use of cosmetics made with insect-based pigment, new inventions for manufacturing insect products, and the availability of new insect-pollinated foods from the Americas.

One of the first things to come to mind, both for historians and entomologists, regarding insects and their impact on humans may well be the Black Death of the mid-fourteenth century. This pandemic, like previous pandemics of plague before it (the earliest being the "plague of Justinian" in the sixth century), has been attributed to the disease caused by the bacterium *Yersinia pestis* and transmitted by fleas from infected rodents to humans.[4] This bacterium causes plague in three forms: bubonic (the most common), pneumonic, and septicemic. While transmission by fleas can result in any of the three forms, the pneumonic form can also be airborne and therefore more rapidly transmitted. Further, bubonic plague can develop into either of the other forms. There has been considerable discussion of what the term "plague" meant during the Renaissance, given that there was no medical understanding of germ theory, nor of insects as vectors (Nutton 2008). Writings of the time about the "pestilence" of 1346–1353 (now referred to as the Black Death) and of recurrences of plague through the sixteenth century focused on describing symptoms, which generally correspond to symptoms of the disease today. However, it is certainly possible that anthrax, tuberculosis, and other diseases may have contributed to pestilential outbreaks. This is an area of ongoing research and discussion among biologists and historians alike, and climate, topography, distribution of rodents, and other ecological and epidemiological factors must be taken into consideration. That said, the identification of pathogen DNA in human remains from the mid-fourteenth century and later confirm that *Yersinia pestis* was the major cause of the Black Death and recurring outbreaks, and that co-infections did exist.

The wave of plague that swept Europe in the late 1340s (and periodically thereafter) had a severe impact on both rural and urban populations. The introduction of plague to Europe was described by Gabriele de Mussis (1280–1356), a lawyer and notary living in Piacenza, in his manuscript *Historia de Morbo*. His account describes the arrival of plague in Italy as a consequence of the siege of the Christian city of Caffa in 1346 by the Tartars. This is also purportedly the first record of the use of biological warfare, although that claim by de Mussis is not undisputed (Benedictow 2004: 52–3). In his account, Tartar attackers besieging the city were struck by plague, dying "as soon as the signs of disease appeared on their bodies: swellings in the armpit or groin caused by coagulating humours, followed by a putrid fever." They then "ordered corpses to be placed in catapults and lobbed into the city in the hope that the intolerable stench would kill everyone inside" (trans. Horrox 1994: 17). As the Tartars abandoned the siege, some of those in the city who escaped returned to their home cities of Genoa and Venice by boat, bringing the disease with them. From those ports it spread throughout Europe; de Mussis also wrote that the plague afflicted "the Chinese, Indians, Persians, Medes, Kurds, Armenians, Cilicians, Georgians, Mesopotamians, Nubians, Ethiopians, Turks, Egyptians, Arabs, Saracens and Greeks" (18).

One of the most often-cited descriptions of the plague in Florence was given by Giovanni Boccaccio in *The Decameron* (1349–1353, trans. Payne 1886):

> I say, then, that the years [of the era] of the fruitful Incarnation of the Son of God had attained to the number of one thousand three hundred and forty-eight, when into the notable city of Florence, fair over every other of Italy,

> there came the death-dealing pestilence, which, through the operation of the heavenly bodies or of our own iniquitous dealings, being sent down upon mankind for our correction by the just wrath of God, had some years before appeared in the parts of the East and after having bereft these latter of an innumerable number of inhabitants, extending without cease from one place to another, had now unhappily spread towards the West. And thereagainst no wisdom availing nor human foresight (whereby the city was purged of many impurities by officers deputed to that end and it was forbidden unto any sick person to enter therein and many were the counsels given for the preservation of health) nor yet humble supplications, not once but many times both in ordered processions and on other wise made unto God by devout persons,—about the coming in of the Spring of the aforesaid year, it began on horrible and miraculous wise to show forth its dolorous effects. Yet not as it had done in the East, where, if any bled at the nose, it was a manifest sign of inevitable death; nay, but in men and women alike there appeared, at the beginning of the malady, certain swellings, either on the groin or under the armpits, whereof some waxed of the bigness of a common apple, others like unto an egg, some more and some less, and these the vulgar named plague-boils. From these two parts the aforesaid death-bearing plague-boils proceeded, in brief space, to appear and come indifferently in every part of the body; wherefrom, after awhile, the fashion of the contagion began to change into black or livid blotches, which showed themselves in many [first] on the arms and about the thighs and [after spread to] every other part of the person, in some large and sparse and in others small and thick-sown; and like as the plague-boils had been first (and yet were) a very certain token of coming death, even so were these for every one to whom they came.

Other descriptions are similar, most including mention of buboes or tumors in the armpit or groin, abscesses, dark blotches on the skin, and intense pain.

The spread of the mid-fourteenth century pestilence was attributed to a variety of causes. Two of the most "scientific" notions were the beliefs that an imbalance of humors in the body led to the disease, and that astrological conjunctions of stars and planets resulted in plague (Horrox 1994; Nohl 2006; and others). "The report of the Paris medical faculty, October 1348 states 'We say that the distant and first cause of this pestilence was and is the configuration of the heavens'" (Horrox 1994: 159). The Parisian physicians then go on to describe the "corruption of the air" brought about by the conjunction of Mars, Jupiter, and Saturn.

Another popular idea was that plague was associated with snakes and other reptiles, as described by Louis Sanctus of Beringen in 1348. Medical treatment included snake venom and meat combined in a mixture called theriac, to be eaten by or rubbed on the patient (Cantor 2001: 172–7). Foul air (miasma) associated with decomposing corpses, swamps, reptile breath, and even volcanic eruptions and earthquakes was widely seen as another source of plague. That "miasma theory" was subscribed to by physicians and the public alike, and elaborate steps were taken to neutralize or prevent the inhalation of foul-smelling air. In Germany, Paracelsus and his followers attributed disease to a disruption in a person's "vital spirit," and relied on "mummies" made of blood or excrement to restore a "magnetic" balance to the vital spirit, an idea somewhat similar to the idea of a balance of humors in the body (Nohl 2006: 79–80). None of these notions included insects as vectors of disease.

Plague struck indiscriminately, killing rich and poor alike. Physicians recognized plague as being contagious, although most subscribed to the miasma theory. Knowing that, many refused to treat patients. Those who did treat patients frequently died (Nohl 2006: 72–6). Several prominent physicians directed that plague victims be dissected at a number of universities in Italy and France, which aided the study of human anatomy but did nothing to find a cure for plague (Nohl 2006: 75–6). Magicians and quacks offering various protections and cures became far more popular than physicians.

Many held the view that sin had brought down the wrath of God in the form of plague. In some cases sin meant iniquity of the people, and in others, malfeasance and corruption in the Church. Church-imposed penances, processions and other rituals, and "plague prayers" were ineffective at combating the plague, and some people, desperate for answers, turned away from the Church and looked for others to blame. Pogroms against Jews, who were believed to have poisoned wells and water sources, occurred all across Europe with many Jews being tortured and burned. Still others, religious fanatics, became flagellants, whipping themselves, "each other, and bystanders in the streets and causing general mayhem" (Cantor 2001: 157). The persistence of disease in spite of actions to repent and to appease the wrath of God had a lasting effect on personal religious belief and belief in the efficacy of Church rituals, practices, and dogma, as well as on moral behavior (or abandonment thereof during the plague years). People turned to magic in the form of amulets and potions to ward off plague. There were social implications as well; notions of kindness and caring for others vanished as people saw caregivers, including family members, clergy, and others succumb to the disease.

The mortality in the mid-fourteenth century was catastrophic. Many scholars have put the mortality rate at 30 to 60 percent for the plague in Europe, drawing on documents including church and abbey records, tax rolls, manor records, and lists of the dead (Gasquet 1908; Zeigler 1969; Gottfried 1983; Benedictow 2004; Nohl 2006; and others).[5] William of Dene, a monk in the monastery of Rochester who chronicled events in Kent, England, wrote:

> To our great grief, the plague carried off so vast a multitude of people of both sexes that nobody could be found who would bear the corpses to the grave. Men and women carried their own children on their shoulders to the church and threw them into a common pit. From these pits such an appalling stench was given off that scarcely anyone dared even to walk beside the cemeteries. There was so marked a deficiency of labourers and workmen of every kind at this period that more than a third of the land in the whole realm was let lie idle.
>
> (qtd. in Zeigler 1969: 162)[6]

As plague devastated Italy in 1350, Pope Clement VI gave indulgences and general absolution to all who might die on the roads to Rome to celebrate the Holy Year. Over a million people traveled to Rome to pray, do penance, and make offerings during Eastertide. Of those, only about one-tenth returned home. Pope Clement VI meanwhile had purchased a home at Avignon and sequestered himself there. However, plague struck Avignon as well, and in order to dispose of the unmanageable number of corpses, the Pope consecrated the Rhone River so that the corpses could be sunk (Nohl 2006). Many other accounts of deaths attributed to plague have been published (see translations in Horrox 1994), not only for the Black Death but also for subsequent plague "visitations" over the next four centuries. Gottfried (1983: 129–33) describes recurring plague cycles every four to twelve years throughout Europe from the mid-fourteenth through the mid-fifteenth century.

Mortality from the Black Death in eastern Europe, Asia, and the Middle East was also high where outbreaks occurred, as described by Hecker (1859: 21):

> Kairo [*sic*] lost daily, when the plague was raging with its greatest violence, from 10 to 15,000; being as many as, in modern times, great plagues have carried off during their whole course. In China, more than thirteen millions are said to have died; ... India was depopulated. Tartary, the Tartar kingdom of Kaptschak, Mesopotamia, Syria, Armenia, were covered with dead bodies—the Kurds fled in vain to the mountains. In Caramania and Caesarea, none were left alive. On the roads,—in the camps,—in the caravansaries,—unburied bodies alone were seen ... In Aleppo, 500 died daily; 22,000 people, and most of the animals, were carried off in Gaza within six weeks.

While that account may have been somewhat exaggerated and some of those numbers have been disputed (Slavin 2019), it is clear that the pandemic had widespread consequences (Spyrou et al. 2019).

Other factors are thought to have contributed to the mortality as well. The recurring catastrophe of plague in the fourteenth century followed a population increase in the thirteenth century, when roughly 90 percent of the population was rural. However, famine had become widespread as the late thirteenth and early fourteenth centuries were subjected to cold, wet conditions (now referred to as part of the Little Ice Age, described above). Those conditions resulted in crop losses and food shortages in the years prior to the Black Death, and it has been suggested that undernourishment and the persistent cold and dampness impacted the ability of people to fight off infections. Following the Black Death and recurring plague outbreaks, the population was severely reduced, and populations in Europe did not recover until well into the fifteenth or sixteenth century (Horrox 1994: 12).

The ability to farm the land was reduced even more as many who tilled the land died, and many survivors were too weak to do so. The demographic shift that resulted caused wages to increase, and a working class (peasantry) developed across much of Europe. In England, the shortage of workers to farm the land led to demands for higher wages. That in turn caused King Edward III to restrict wages and to fine or arrest those workers who in defiance of his orders demanded (and got) higher wages. Further, the King also fined or arrested those who paid their workers higher wages. Knighton, a canon of Leicester, described the situation:

> The King sent into each county of the kingdom orders that harvesters and other workmen should not obtain more than they were wont to have, under penalties laid down in the state made for the purpose. But labourers were so elated and contentious that they did not pay attention to the command of the King; and if anyone wanted to hire them he was forced to pay them what was asked, and so he had his choice either to lose his harvest and crops or give in to the proud and covetous desire of the workmen. When this became known to the King he levied heavy fines upon the abbots, priors, and the higher and lesser lords, as well as upon the greater and smaller landowners in the country, because they had not obeyed his orders, and had given higher wages to their labourers ....
>
> (qtd. in Gasquet 1908: 230)

The growing conflict between these survivors of plague and their lords, building on the unrest within the system of serfdom that preceded the Black Death, culminated in various uprisings such as the Peasants' Revolt in 1381 in England. Gasquet (1908: 232) wrote that "... by the beginning of the next century the whole tenure of land had been changed in England by the great mortality of 1349 ..." and further, "... although for the next thirty years the lords and landowners fought against the complete overthrow of the mediaeval system of serfdom, from the year of the great mortality its fall was inevitable ...".

Similar demands for higher wages were heard across western Europe, with similar consequences: the rise of a new socioeconomic system in which the bargaining power of peasants and craftspeople increased. Gottfried (1983) described the shift from serfdom to a system whereby peasants worked the land and paid rent; both peasant and landlord had a copy of the tenant agreement, hence the system was called "copyhold." (He also noted that this did not occur across much of eastern Europe, where serfdom persisted into the nineteenth century.) Benedictow (2004: 360–2, 388–9) identifies the socioeconomic changes that took place as a turning point, resulting in dramatically improved living conditions for workers and peasants over the century following the Black Death. In urban areas, guilds of craftsmen (and in some places, women) arose and became powerful as skilled workers came into demand. Guilds of silk workers and dyers were among the most powerful. Guilds gained economic and political prominence in many cities across Europe, and they became an important part of the fabric of European society. One might conclude that if anything positive came from the plague which ravaged Europe from the mid-fourteenth century onward (and was at least in part vectored by fleas), it was the promotion of that shift from serfdom to a working class with more economic freedom. That increased economic prosperity in turn gave rise to unprecedented economic growth and the rise of banking and industry, fostering technological innovation. The wealth that accrued to the ruling class, including the Medici family in Florence, allowed for patronage and the promotion of a climate of intellectual and artistic exploration and expression.

The mortality associated with the Black Death and subsequent recurrences of plague clearly impacted food supplies as the number of people producing food decreased. However, while this impact was an indirect effect of the insect vectors that spread disease, insects also had more direct effects on food production. Although not much was written concerning insects that may have damaged crops on a day-to-day basis, there are records of locust swarms attacking crops and destroying vegetables. Camuffo and Enzi (1991) compiled records of locust outbreaks in Italy and elsewhere in Europe during medieval and Renaissance periods (and through 1800), noting that the sixteenth century had the greatest number of outbreaks. The descriptions of that time generally describe large swarms, and regard them as a "wonder," "great misfortune," or other type of unusual event. Similar swarms of locusts were recorded in Africa by European missionaries in the mid-sixteenth century. Such swarms were known since biblical times and in fact occur today; the desert locust, *Locusta migratoria*, is the most common species forming large swarms. In some parts of Europe, locusts were brought before ecclesiastical courts (as were other animals that were seen as causing harm), and were commonly excommunicated. The theological implications were complicated, but this response by the Church was often seen as successful when the locusts moved on (as they would do, excommunication or not).

Although records of insect damage to food, especially by swarms of locusts, show an awareness of the negative impact of insects on agriculture, there were certainly positive impacts as well. The European honeybee, *Apis mellifera*, had long been intentionally raised for its products; honeybees continued to be important providers of honey and wax, and the products derived from them, throughout the Renaissance. The role of bees in crop pollination was undoubtedly also important, although it may not have been recognized. New techniques for handling bees were also developed during the Renaissance, and treatises on apiculture appeared. The invention of the printing press made these techniques and information on apiculture widely accessible.

Silk cloth had been a highly prized commodity since Roman times, and although wild silks were well known (Peigler 1993), the finer-textured silk from the silk moth *Bombyx mori* was the most commercially valuable. While silk fabrics were imported from the East, raw silk was also imported from China and manufactured into cloth locally. Artisans in Europe, and especially in Italy and France, became masters at reeling and "throwing" silk (twisting into threads or yarns), as well as dyeing and weaving fine silk textiles.[7] English royal inventory records show, for example, a luxury striped silk from Paris as having been purchased for King Edward II of England in 1332–1333, providing evidence of a Parisian silk industry at that time. The silk industry grew dramatically in urban centers such as Milan, Florence, Venice, Paris, and other cities in the fourteenth century; Molà (2003), Tognetti (2005), and Farmer (2014) provide comprehensive and detailed accounts of the development of the silk industries in Venice, Florence, and Paris respectively. Accomplished silk workers were in demand, and guilds of silk craftspeople became very powerful, setting prices and regulating the production of silk at every step in the process. Silk, always in high demand among the wealthy, had become a "signature" trade item of these and other (mainly Italian) cities by the late fourteenth century. Skilled artisans were given special privileges and/or were prohibited from moving to other cities and taking their skills with them. Silk production and the techniques involved were unique to each city in which the silk was made, and they became highly guarded trade secrets; violators were harshly punished.

The silk industry became a major economic driver in these cities during the fifteenth century, which became the hubs of trade networks that developed across Europe and the Mediterranean basin. Manufacture of silk provided both skilled and unskilled workers with jobs, and the profits gained fueled the success of a merchant-banking industry as well as an increased source of tax revenue for the aristocracy and revenue for the Church. It also led to technological innovations and entrepreneurship, giving rise to new industries such as the manufacture of hydraulic spinning machines.[8] The system of patronage that gave rise to the proliferation of commissioned art, including Michelangelo's paintings and sculptures, Raphael's work, and sets of magnificent silk tapestries, had its economic roots in the silk industry that flourished in Italy and elsewhere. Other industries such as shipbuilding contributed to the economic growth that spurred the Renaissance, but the economic role played by the silk industry cannot be overstated.

Raw silk itself was largely obtained from Eastern sources through the mid-fifteenth century. However, raw silk became much more expensive following the Ottoman Empire's cessation of trade with China, which effectively shut down many trade routes. Although silkworms had been raised in Italy and Greece since the time of Justinian, the difficulty in obtaining sufficient quantities of raw silk and rising prices led a number of European rulers to actively promote sericulture. For example, in France, De Quatrefages (1873) reported the planting of four million mulberry seedlings under King Henry IV, and in mid-fifteenth-century Tuscany each peasant was required to plant at least five mulberry trees. Sericulture flourished in France and Italy during the fifteenth and sixteenth centuries, providing additional income for peasants who could sell the raw silk to merchants and guilds of silk workers in the cities. Like apiculture, sericulture also saw the benefit of the printing press; books and pamphlets on various techniques for raising silkworms were widely distributed.

Silkworms were not the only insects that had a major impact on the economy of textile production. Silks, along with cotton, wool, and other fibers were dyed in a variety of colors, many of which were derived from plants and minerals. The reds, however, came primarily from scale insects, long known as a source of red pigments. A number of different scale insect species, including kermes, lac, and cochineal scales, were used to produce dyes of various shades and properties. Many of those scales were raised in eastern European countries and imported into the western European centers of textile production. The pigments obtained were also used in inks and paints, as well as in early cosmetics. Recipes for all of these pigments and pigment products were often guarded in the same way that silk artisans guarded their methods. Guilds of dyers were a critical part of the silk industry as well as other textile industries.

Scale insects from the New World, however, became the source of choice for the most intense and beautiful reds.[9] Following the colonization of Mexico by the Spanish, the cochineal scales that had been raised there for centuries by the indigenous peoples were in great demand. Although hard to imagine, literally tons of scale insects were sent back to Spain by the Spaniards, and these insects were a source of wealth for those who controlled their trade in Europe. The red silks that were produced using these scales were highly sought after, and in fact the Catholic Church officially adopted red as the color to be worn by its Cardinals. Modern chemical analysis has shown that the New World scale insects were the source of red pigments used in various fabrics, as well as in paintings by many of the Renaissance masters (see Padilla and Anderson 2015). Records of Spanish administrators in the New World show that these scales were intentionally raised and cared for by the Indigenous peoples of what is now Mexico, and modern genomic analysis has suggested that strains of cochineal scales were selected for (Van Dam et al. 2015). Cochineal scales brought back to Europe from New Spain became an important part of the European economy.

Aristotle's *Historia animalium* and Pliny the Elder's *Naturalis historia*, along with other ancient texts, formed the basis of formal authoritative knowledge about the natural world, including what was known about insects, at the beginning of the Renaissance. Interest in the natural world increased dramatically during the fourteenth and fifteenth centuries with the rise of a secular humanist intellectual climate that began in Italy, and as Ogilvie (2008b: 1) notes, "natural history was invented in the Renaissance."[10] Much of this interest took the form of efforts to discover, or rediscover, what the ancient classical texts said about various species of plants and animals. Of special concern were the medicinal and other uses described in these texts, but they were also considered authoritative for descriptions of known species of animals and plants. Those texts defined ways in which many insects were regarded; for example, the notions that flies and worms (maggots) were generated in filth, that a king ruled a hive of bees, and that insects have no blood were accepted as facts. The classical designations of "worms" and "imperfect creatures" were broadly inclusive and placed many insects together with other arthropods, segmented worms, leeches, and sometimes even snakes and frogs. Scholars in the fourteenth and fifteenth centuries began to record observations of the natural world, including insects, and to describe nature in more detailed and concrete terms, often accompanied by illustrations.

In the fourteenth and fifteenth centuries, this expanding interest in the natural world was supported by Italian patrons such as Duke Federico of Montrefeltro, Cardinal Bessarion of Venice, and others who founded colleges and libraries. They also sponsored the collection, translation into Latin, and dissemination of texts in many disciplines. This intellectual climate gave rise to the avid collection of specimens, which were then included in encyclopedic works that attempted to catalog human knowledge of the natural world in the context of humanist natural philosophy. Often modeled after and drawing heavily on the ancient texts, these encyclopedias incorporated what was then current knowledge of species and phenomena, often with a practical or medicinal focus. For example, Giorgio Valla of Placentia (1447–1500) compiled *De expetendis et fugiendis rebus* (published in 1501), an encyclopedia of forty-nine books, eighteen of which described aspects of the natural world and seven of which focused on medicine. He did not deal with species per se but did describe the harmful effects of venomous animals, including bees. A contemporary of Valla's,

Polydore Vergil, published a compendium also dealing with the practical uses of animals and plants.

In the sixteenth century, as new organisms were being discovered in the New World, the cataloging and description of species became the focus of compendia and comprehensive encyclopedic works such as Conrad Gessner's *Historia animalium* (1551–1558 and 1587), and Thomas Moffett's *Theater of Insects* (1658). These efforts reflected the increasing interest in the natural world, curiosity about that world, and an intent to learn more about living organisms for practical, philosophical, and aesthetic reasons. The natural philosophy of the early Renaissance blossomed into scientific inquiry and what we now refer to as "natural history" in the fifteenth and sixteenth centuries and later. Contributing to that natural history was the emerging practice of collecting and preserving specimens, and extensive insect collections as well as herbarium collections and collections of specimens of a variety of animals were made. Individual scholars shared specimens and information, and the acquisition of knowledge about the natural world became, in many cases, a collaborative effort. That practice of collaboration became a standard practice in scientific inquiry that persists today.

Insects were not initially the focus of interest in the worlds of medicine (with the exception of some venomous members of the group of "imperfect creatures"), art, and philosophy, where more attention was often given to plants and larger animals. Insects came to play an increasingly important role as the Renaissance progressed, partly because of the increasing interest in the natural world, especially after the voyage of Columbus in 1492, and partly because improving technology enabled more detailed and sophisticated representations of insects in a variety of mediums. Further, the invention of the printing press in the mid-fifteenth century made possible the dissemination of information about insects, with images, in publications from pamphlets on sericulture and beekeeping to encyclopedic compendia. These compendia attempted to bring together all that was known—about medicine, about plants and animals, about non-living components of the natural world such as rocks and fossils, and for the first time in the sixteenth century, about insects. Because images of insects, "visual understanding," was an essential part of those works, the evolution of insect depiction merits consideration. Indeed, in the Renaissance as well as today, "a picture is worth a thousand words."

Long recognized as objects of beauty, some insects, such as butterflies, had been incorporated as decorative additions to illuminated manuscripts during the medieval period, and this continued into the fourteenth century. These images were stylized and not meant to portray realistic insects, but to be purely decorative. That changed in the fifteenth and sixteenth centuries; drawings of insects became more detailed (as did drawings of other organisms). Still used to illuminate some manuscripts, more realistic images of insects were meant to illustrate descriptions of those insects in compendia, as well as other literary works. One such representation of insects appeared as a woodcut in von Megenburg's *Buch der Natur* in 1475, and serves to show insects that were considered important at that time (Neri 2011). Still stylized, it shows bees and beehives, flies, grasshoppers, ants in an anthill, and a butterfly on a flower, along with a spider and some other creatures. Relative size and numbers of the insects reflect their importance.

The representation of insects became much more detailed and realistic in the sixteenth (and seventeenth) centuries. Artists began to include insects (and other arthropods) in depictions of a variety of subjects, mostly religious or having religious overtones, in paintings, drawings, and tapestries. The art that grew out of the experience of the Black Death and other episodes of plague often showed death, plague victims, and flies, which were associated with death and decay. Some of those associations were based on Aristotle's descriptions of certain insects such as maggots (worms) and flies arising from filth. Others were based on the insects that would ordinarily be associated with bodies, including victims of plague, and those insects, especially flies, became symbols of death. The fly appears in the fifteenth-century painting *Madonna and Child* by Crivelli, and has been interpreted as representing Christ's death. The artist Hieronymus Bosch in his early sixteenth-century painting *The Origins of Sin* depicts fallen angels transforming into insects as they fall.

Insects took on other religious and symbolic meanings as well, more beneficent in nature. Butterflies and scarab beetles, for example, were associated with resurrection. Lady beetles, which eat harmful aphids on crops, were considered gifts from the Virgin Mary, hence their name (Kritsky and Cherry 2000). The scarab beetle had long been a symbol for Christ, having been described thus by Ambrose of Milan in the fourth century, and the persistence of this symbolism was reaffirmed by Joris Hoefnagel and the Jesuit polymath Athanasius Kircher in the sixteenth and early seventeenth centuries respectively. The symbolic meanings attached to particular insects were not lost on Renaissance writers, who used insects as metaphors for human virtues, as well as vices.

As artists turned from strictly religious subjects to secular subjects, including more realistic and instructive representations of nature, insects were included in depictions of landscapes, gardens, and other places and events. The insects became an important part of these visual descriptions, but were not central to the artwork. Albrecht Dürer was the first to make an insect the focal point—indeed the only subject—of his 1505 drawing of a stag beetle, initiating what Neri (2011) calls "specimen logic," or focus on individual specimens, as an approach to depicting insects that persisted throughout the sixteenth century and beyond.

Other notable artists followed Dürer's lead, and one of the most influential, both in the world of art and in a nascent world of entomology, was the

Flemish artist Joris Hoefnagel. Although he drew a number of imaginary and fanciful insects, he also drew exquisitely detailed actual insects as illustrations, showing the importance of careful observation of detail as well as accuracy in depicting that detail. In his sixteenth-century work *Four Elements*, he drew morphologically accurate insects, in some cases with the plants on which they were found, and he also provided annotations and inscriptions that showed knowledge of his subjects (Neri 2011; Bass 2019). Neri (2011: 22) summarizes the significance of Hoefnagel's work:

> Hoefnagel's manuscript illustrations and published engravings provided a corpus of insect imagery that could be used as models and that helped to establish a framework for understanding insects as prized and visually complex subjects and objects. In fact, these images were crucial tools for artists and other practitioners engaged in developing the emerging concept of "the insect world."

It is worth noting that in addition to producing beautiful images of insects (and other animals), Hoefnagel also drew "pseudoinsects," part one insect and part another but blended so well that unless one recognized the parts, one would think the "insect" was real. These imaginary creations were used decoratively in various manuscripts. Hoefnagel and many other artists of his acquaintance used images to record events and places, document the history of art in the Netherlands, express personal friendships, and give silent voice to political and religious views. Insects were frequently used in these works, and in some cases seem to take on cryptic and symbolic meaning in times of political tumult.[11] Hoefnagel also did commissioned work for the Catholic Church, and so contributed to the incorporation of insects into religious art.

In the emergence of active efforts to describe all aspects of the natural world, both collections and depictions of insects and other creatures took on a central role in shaping the study of nature. The collecting of insects, as well as plants, shells, fossils, animals, and birds, was widespread in the sixteenth century, and specimen cabinets were a popular way to display these objects. Indeed, Ulisse Aldrovandi had what amounted to a private museum in Bologna, and other collectors who would eventually catalog and publish descriptions of insects (and plants) included Thomas Penny, Thomas Moffet, and Conrad Gessner. Specimens of insects (as well as botanical specimens) were often exchanged, and extensive collections were visited by many scholars of natural history (Neri 2011). Artists were hired to produce illustrations of these specimens; illustrations could be more easily circulated than the specimens themselves. These scholars' efforts to collect and describe insects spanned the last half of the sixteenth century and involved the work of many individuals. The results were the first comprehensive publications on insects, Aldrovandi's *De animalibus insectis* (1602) and Moffet's *Theatrum insectorum* (*Theater of Insects*) (1634), the latter being a collaboration including work of Edward Wotton (from his *De differentis animalium* (1552)) and work done by Conrad Gessner, Thomas Penny, and Moffet himself.[12] These works used images of the insects along with descriptions (and other text), and the fact that they could be published and circulated made the study of insects widely accessible. These illustrated compendia of insects (and other compendia covering other animals and plants) attest to the increasing interest in studying the natural world. These works also constitute the beginning of what might be considered early biology, natural history, and entomological studies.

Records of insects discovered in the New World were also kept by explorers, priests, and civil administrators, primarily for the purpose of describing colonial regions, economic uses, food sources, and health. For example, the *Histoire naturelle des Indes* (the *Drake Manuscript*) contains drawings of insects, albeit not lifelike representations, in their habitats. The sixteenth-century *Matrícula de tributos* (Tribute List) from central Mexico shows in pictographs the cochineal scale insects paid as tribute to the Aztec emperor. Bernardino de Sahagún's *Florentine Codex* (1575–1577) describes the cultivation of cochineal scale insects and their hostplant (Kerpel 2015). In the British colonies, John White, governor of the Roanoke settlement and an artist as well, made drawings of many of the plants and animals, including some insects, and included some comments concerning them; his drawings were published as a series of engravings titled *America* in 1590 (Neri 2011). Such drawings and descriptions of insects tended to be more practical in nature, conveying useful information about the New World to those in the Old World, but not focused on the insects themselves.

The confluence of the efforts to describe the natural world, not as a means to glorify God but as an end in itself, and the discovery of the New World with a completely different and largely unknown flora and fauna led to the beginnings of a scientific approach to natural history. Indeed, Hoefnagel's *Four Elements* is an exemplar of early scientific inquiry, a purposeful effort not only to describe nature but to understand it. His exquisite images and written inscriptions on the sheets of each folio, while drawing on other more encyclopedic works such as Gessner's *History of Animals*, are unique in that, unlike Gessner's and other works, they were not intended to be widely available for "the unlocking of nature's manifold secrets for Humankind" (Bass 2019: 192).[13] Along with the encyclopedic descriptions of nature made by his contemporaries, Hoefnagel's work helped to set the stage for the field of natural history to flourish in the seventeenth through the nineteenth centuries. Organized efforts to discover, describe, and understand the ways in which organisms interacted in nature eventually culminated in the works of early entomologists and natural historians such as Henry Bates, Alfred Russell Wallace, and Charles Darwin (who was an avid beetle collector) in the nineteenth century.

Insects also found their way into literature, having already been a part of classical literature where they were often used to represent various aspects of human nature, as in Aristophanes' play *The Wasp*. More familiar are the fables of Aesop, such as "The Ant and the Grasshopper," where various insects were used to teach moral lessons. (That stories such as this have endured through time is evidenced in the making of Disney's 1934 animated film *The Grasshopper and the Ant*, and more recently, Pixar's 1998 animation *A Bug's Life*.) During the Renaissance, insects appear in the works of Chaucer, such as *In the House of Fame* where bees are used as an analogy for the sound made by a group of people. In other works, Chaucer's references are not to specific insects but to "worms," a category that at that time included a host of insects as well as other invertebrates.

Shakespeare also incorporated a number of insects into his plays; his use of not only insects but other animals, plants, and elements of natural history was analyzed by Robert Patterson in 1838. Regarding his analysis of Shakespeare's use of insects:

> According to Patterson's study, Shakespeare covered seven insect orders: Coleoptera (beetles), Hymenoptera (bees, wasps and ants), Diptera (flies), Lepidoptera (moths and butterflies), Orthoptera (grasshoppers and crickets), Aphaniptera (fleas, now called Siphonaptera) and Aptera (lice, now called Phthiraptera). Patterson was clearly up with the taxonomic times. Grasshoppers and crickets had only just been separated from the Hemiptera (cicadas and planthoppers) in the late 1700s, but some of Patterson's contemporary authors were still referring to them as hemipterans.
>
> (Saunders 2016)

Bees and wasps, stinging insects, were used as metaphors for conflict or punishment in several of his plays. In others, he uses bees as givers of honey, and butterflies as representing beauty. Dyer (2010), in his 1883 volume *Folk-lore of Shakespeare*, provides specific references to plays, including information from Patterson's earlier work; he introduces his chapter on Insects and Reptiles with commentary on the accuracy of the Bard's observations of nature:

> As Dr Johnson has truly remarked, Shakespeare is "the poet of nature," for "his attention was not confined to the actions of men; he was an exact surveyor of the inanimate world; his descriptions have always some peculiarity, gathered by contemplating things as they really exist. Whether life or nature be his subject, Shakespeare shows plainly that he has seen with his own eyes." So, too, he was in the habit of taking minute observation of the popular notions relating to natural history, so many of which he has

> introduced into his plays—using them to no small advantage. In numerous cases, also, the peculiarities of certain natural objects have furnished the poet with many excellent metaphors.
>
> (Dyer 2010: 236)

One of the most famous literary works in which an insect was mentioned is the poem "The Flea" by John Donne (2011). Although not published until 1633 after Donne's death, most scholars think it was written in the 1590s. In that poem, the flea becomes a metaphor for the proposed union of two lovers, the flea having drunk the blood of both and so united their blood. The use of such metaphors and analogies suggests that all of these authors had a clear sense of the behaviors, habits, and value of these insects in the context of Renaissance society.

Through the centuries of the Renaissance, insects moved from the periphery of knowledge to becoming one of several central foci of inquiry into the natural world. These foci would mature into fields of scientific study: biology, botany, anatomy and physiology, vertebrate and invertebrate zoology, and entomology. Connections of insects with human culture became more numerous and complex as knowledge of insects increased, especially as new cultures and new insects (and other organisms) were discovered in the sixteenth century. At the same time, the roles played by insects in disease transmission would remain a mystery until the nineteenth century.

It is against the backdrop of change that the many roles of insects in human culture are here examined. This volume explores eight categories of cultural interactions with insects, using historical, physical, visual, and archaeological evidence, as well as drawing on the excellent scholarship that has emerged in recent decades on various aspects of insects, their natural histories, and their products. We address not only the importance of insects to human health during the fourteenth to sixteenth centuries, but also significant contributions of insects to human society and thought in a practical sense, an artistic sense, an intellectual sense and a scientific sense. Leonardo da Vinci saw the universality of connections, and threads of those connections run throughout these chapters as well. Our hope is that this volume brings a greater awareness and understanding of the impact of insects on human cultures across the Old and New Worlds during the centuries termed the Renaissance.

CHAPTER ONE

Insect Knowledge

PETER C. MANCALL

The mosquitoes were everywhere. That, at least, is how an unknown European artist depicted a swarm somewhere in the greater Caribbean basin in the waning years of the sixteenth century (Figure 1.1). A French writer, who might or might not be the same person who created the watercolor of the mosquitoes, offered a description. "They are small flies which are so small that one cannot see them, they are very dangerous," the surviving manuscript reported. "When there is no wind and the weather is calm, they come in droves attacking people, stinging them in such a manner that one would take them for lepers." The writer emphasized the human pain that soon followed. "Where they bite," the author noted, "the flesh swells up like a pea and if one kills said flies where they have stung, this protects them from the swelling." The European, perhaps thousands of miles from home, reported something else: the locals knew what to do. "The Indians," the text noted, "make a fire in their houses in order to keep them away and they only appear at night, retreating during the day close to the sea in the sand" (*Drake Manuscript*, f. 72, p. 263).

The presence of the mosquitoes in this manuscript comes as no surprise. While the artist or artists—modern scholars have detected two different styles in the watercolors—remain unknown, efforts to depict foreign places for an audience at home were not novel. Ever since Europeans began to travel beyond the edges of their world, they wrote accounts (and sometimes added pictures) of what they saw. They did so by reporting details as accurately as possible, or at least that was the impression that they wanted to convey even if some of their reports, like that of the fabled English knight Sir John Mandeville, contained utter nonsense and fantastic tropes about the so-called monstrous

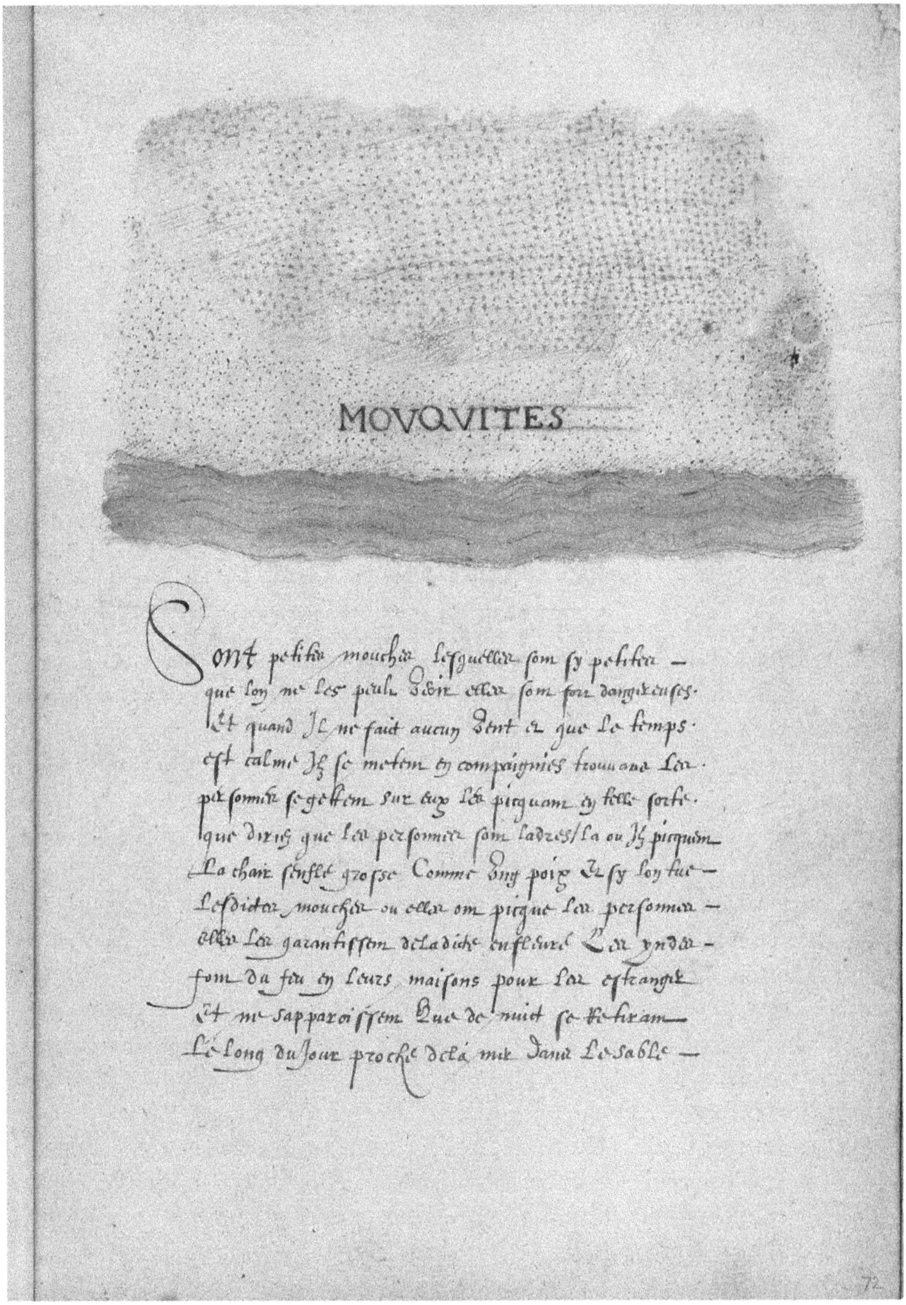

FIGURE 1.1 An unknown artist captured the sensation of a mosquito swarm in the *Histoire Naturelle* even if no one at the time had the capacity to illustrate the insect with more precision. (Reproduced with permission of The Morgan Library & Museum. MA 3900. Bequest of Clara S. Peck, 1983.)

races that lived somewhere beyond the edges of the world. (The *mappa mundi* in the Hereford Cathedral in England, painted on parchment at the start of the fourteenth century, took this practice to a logical extreme by placing monsters including blemmyes with their heads on their chests and sciapods with giant feet to shield them from the sun at the far edges of the map [Kline 2001: 141–64]). Most travelers likely understood that conveying false reports would not burnish their reputations or secure them future support. So instead of making things up out of their imaginations, they instead tried to describe what they saw with the analytical perspectives they already had, typically by comparing the new with the known and employing analogous logic to explain distant phenomena to people who would likely never encounter them. The artist and author of the short report about the mosquitoes did just that: the writer of the caption likened this insect to a common fly. This American animal was unusual since it could not be seen and caused pain. But the act of describing it made sense.

The era typically known as the Renaissance brought many changes to European societies. Great advances in art and science, seen in the works of Michelangelo (1475–1564) and Leonardo da Vinci (1452–1519), contributed to changes in the ways that European men and women understood the world around them. Challenges to clerical practices emanating from Rome led to the fracturing of Christendom during the Reformation, a movement that gained momentum when the German theologian Martin Luther (1483–1546) posted his critique of the Church, known as the "Ninety-Five Theses," at All Saints' Church in Wittenberg in 1517. While that movement produced bitter conflicts and outright warfare across parts of Europe, it also contributed to the spread of knowledge. Religious writers publicized their arguments through print, a development that could only take place because of the proliferation of presses in the years following Johann Guttenberg's invention of metal movable type in 1450. (He was not, it should be noted, the first to develop the idea of movable type since the Chinese had their own technology for doing so since the eleventh century CE [Bussotti and Qi 2014].) Still, while the spread of printing presses facilitated the dissemination of information, Europeans continued to rely on manuscripts to transmit knowledge into the eighteenth century (Eisenstein 1980; Love 1993; Johns 1998), including information about the natural world.

Among the beneficiaries of the spread of the printing press were scholars who wanted to demonstrate their knowledge and travelers looking for fame and fortune based on their exploits. Both of these groups, as it turned out, had a deep interest in insects. Their attentiveness made sense given the role that fleas, flies, lice, and mosquitoes played as vectors of lethal diseases in early modern Europe and in European colonies abroad. Later, after 1600, scientists and policymakers would come to understand the relationship between dirty, rat-infested cities and polluted water supplies, on the one hand, and the spread of insect-borne diseases including dysentery, malaria, plague, and typhoid, on the other. Armed with emerging scientific knowledge, they set out to pave streets, improve sewage systems, drain swamps, and set up ways to dispose of garbage (Riley 1986).

But well before the coming of new understanding of how diseases spread in cities, Europeans and others shared knowledge of the insect world. One twelfth-century Chinese scholar compiled information about invertebrates such as snails and centipedes, and insects including butterflies along with other creatures (Fan 2011: 81–8). Native Americans did so also, employing oral history to share knowledge from one generation to the next. When Spaniards arrived in Mexico after the conquest initiated by Hernán Cortés (1485–1547), they quickly set about trying to learn what Indigenous people knew. Other arriving Europeans were keen to record native information too. As a result, Indigenous knowledge survived in texts that contained a wealth of details about the natural world and natives' understanding of it. The French observer who wrote about the mosquitoes understood, like others, that Americans possessed deep understanding of creatures that Europeans could not even see.

INTELLECTUAL ANTECEDENTS

By the start of the fourteenth century, populations around the globe had deep knowledge of the insects that shared their environment. In the ancient Mediterranean, various communities integrated the bodies of insects into their medicines, a practice that continues in much of the world today (Beavis 1988; Costa-Neto 2002). Further, insects figured in the diets of humans in many places, including hunter-gatherers, some of whom continue to share information about the nutritional benefits of insect consumption. The phenomenon was so common and acceptable that the Bible contained rules for eating insects (in Leviticus 11:22) and notes that they were eaten by John the Baptist when he was in the desert (Matthew 3:4), surviving on honey and locusts (Morris 2008; Si and Turpin 2015). Perhaps no insect was as popular as the bee. Ever since ancient times, various communities had cultivated hives as a way to get honey to improve their food and their medicines. During the Renaissance, much knowledge about bees from ancient Greece and Rome re-emerged in texts, paintings, and histories that circulated in Europe. European insect knowledge contained more than traces from the ancient Mediterranean basin. The Old Irish beekeeping tract *Bechbretha* revealed the role that hives played in monastic and other communities on the western edge of Europe (Charles-Edwards and Kelly 1983).

The most systematic body of information from antiquity about insects that survived to the Renaissance was book eleven of Pliny's *Natural History*.

Even in the first century CE, as his text revealed, scholars had become fascinated by the vast range they observed in the bodies of what he called "creatures of immeasurably minute structure" that were "great of number and of many kinds." These entities "have the habits of land-animals and of flying animals, some lacking wings, for instance centipedes, others winged, for instance bees, others of both kinds, for instance ants, some lacking both wings and feet." But Pliny understood that they needed to be understood as a group as "insects" because they shared certain physical traits that revealed "a craftsmanship on the part of Nature that is more remarkable than in any other case." Even still, insects baffled Pliny and his contemporaries. Did they breathe? Did they have blood? Their size invited more questions. Where, he asked, "did Nature find a place in a flea[1] for all the senses?" How could such a minute being see something? Or taste, smell, or hear? Pliny put forth the kinds of questions that would prompt serious scientific exploration during the Renaissance. Again, in reference to the flea, how had nature

> attached the wings, extended the legs that carry the feet, placed a ravenous hollow to serve as a stomach, kindled a greedy thirst for blood and especially human blood[?]. Then with what genius [nature] provided a sharp weapon for piercing the skin, and as if working on a large object, although really it is invisibly minute, created it with alternating skill so as to be at once pointed for digging and tubed for sucking!
>
> (Pliny 1938, III: 433–5)

Pliny's interest extended well beyond the flea. He wrestled with questions relating to insect morphology and practices. He spent more time, as might be expected, on insects that were the most important in the ancient world. Anticipating a trend that would prove powerful during the Renaissance, he singled out the bee "because they alone of this genus have been created for the sake of man. They collect honey, that sweetest and most refined and most health-giving of juices." They were hard workers, built complex structures, "have a government and individual enterprises and collective leaders," and, most surprising, "they have a system of manners that outstrips that of all other animals, although they belong neither to the domesticated nor to the wild class" (Pliny 1938, III: 439–75, quotations at 439). He wrote about the transformation of caterpillars into butterflies and of the utility of their webs to spin silk. He wrote about spiders, though not in as much length as he did about bees because most "need not be described, as they are so well known." He wrote about efforts to suppress locusts, including a law in Cyrene to wage war against them three times each year; any person who shirked duty in this campaign would be sentenced like a deserter. His descriptions fit the information he had available. He believed that fire produced a certain kind of fly and that scorpions and geckos were insects (Pliny 1938, III: 479, 481, 485–9, 497, 505–7). His writings about beetles attracted serious attention into the twentieth century (Scarborough 1977).

However inaccurate such descriptions might seem, Pliny's views, carried through the centuries between antiquity and the Renaissance, constituted the most thorough account of insects available to those who studied insects in Europe. Scholars across the continent copied it well before the advent of movable type (Thorndike 1936; Lagorio 1973). Peter Martyr d'Anghiera (1457–1526), a chronicler of Christopher Columbus, and his intellectual rival Gonzalo Fernández de Oviedo y Valdés (1478–1557), each drew on Pliny in their observations of nature in the early decades of the sixteenth century (Beagon 2014). One scholar after another produced editions of Pliny's texts from at least the fifteenth century, including the renowned Dutch humanist Desiderius Erasmus (1469[?]–1536), who completed his edition in 1525 and proclaimed, like other scholars of his era, that his was the best version (Tomlinson 2012). Specific aspects of Pliny's views on insects surfaced time and again, including his belief in the primacy and social complexity of bees (Woolfson 2010).

Pliny's *Natural History* consistently attracted scholarly attention. There was at least one edition, one historian has noted, created in every century in medieval Europe. The spread of print technology led to a virtual explosion of new copies and editions. The Venetian printer Johannes de Spira (d. 1470) produced an edition in 1469, making it one of the first major books to roll out of a press. Before 1500, there were fifteen printed editions in Latin and three Italian translations. Scholars produced dozens of editions during the sixteenth century, most but not all in Latin. Unlike medieval scholars who were content with reproducing Pliny's evidence, these Renaissance-era translators wrestled with the nature of the text and the desire of each to produce a better version than those that had come earlier. In the process of debating the utility of the *Natural History*, including what role it might play in university education, Renaissance scholars opened a new way of thinking about the natural world. As the historian Charles Nauert, Jr. wrote in 1979, the focus on Pliny and his sources that had led to debates about the accuracy of different editions evolved into a philosophy about how to analyze and contemplate the environment. "The true goal of the natural philosopher," Nauert noted, "was coming to be more clearly conceived as the determination of the truth about nature, not just the determination of what a long-dead author had really written" (Nauert 1979). This shift in perspective was crucial for the understanding of nature—including the vastness of the insect world.

THE EXPANSION OF ENTOMOLOGICAL KNOWLEDGE

In 1300, sensibilities about nature remained in place as they had been for perhaps a thousand years. Human communities had experienced some growth, but even the largest cities in the world would be insignificant compared to the urban areas that had become more numerous and larger by the start of the seventeenth century. The changes began around the turn of the first Christian millennium. In Europe, cities on the Italian peninsula became larger by the start of the eleventh century, a result of increased trade across the Mediterranean. A century later commerce across the North Sea, Baltic Sea, and along the Rhine drove the expansion of cities farther north. In England, London rose to its dominant position by the start of the fifteenth century (van Steesel 2016). The rising population of cities in particular and Europe more generally signaled recovery from the devastations of the plague that had reduced the continent's population by approximately one-third in the late 1340s. (Though the plague had been spread by an insect, Europeans at the time were not aware of the complex relationship between rats, fleas, and human ailment and instead looked for human culprits who must have played a role in the pandemic [Herlihy 1997: 17–38; Cohn 2018: 48–67]). By 1500, Europe had six hundred cities with a population of at least five thousand. A century later, thirty-seven of these urban areas had a population of at least forty thousand (de Vries 1989). Across Europe, the growth in population following the plague was staggering. The population in England, to take one notable example, increased from just over 3 million around 1550 to approximately 4.1 million by 1600 and then to 5.2 million by 1650 (Poos 1989: 799).

Cities in the Americas are more difficult to measure, but the largest of them grew as well during these centuries. The Mexica began to build the city of Tenochtitlán around 1325. By the time the Spanish conquistador Hernán Cortés arrived in 1519, perhaps 200,000 people inhabited a sophisticated city dotted with remarkable institutions, including a substantial aviary and a menagerie, signs of the Aztec emperor Moctezuma II's interest in the natural world. Cuzco in the Andes rose even more quickly after the victory of the Incans over their neighbors in the late 1430s. By the time the conquistador Francisco Pizarro arrived a century later the city had elaborate architecture and spectacular wealth (Smith 2005: 411–12; James-Chakraborty 2014: 16–29). Cahokia, the largest city north of Tenochtitlán, had a population of perhaps ten thousand at its core and another twenty thousand along its periphery around the year 1000. But a series of long-term droughts and a mega-storm in the middle of the twelfth century led many to decide to leave. By 1300, the city was largely deserted, its large mounds remaining as memorials to its stunning rise and fall (Pauketat 2009; White et al. 2019). Population growth was uneven, but in some places the increase was dramatic. The population of China likely doubled from *c.* 1400 to *c.* 1600 (Durand 1977: 263).

In all places, larger cities supported increasingly sophisticated cultural institutions, including increased attention to the natural world and its myriad denizens. Even still, while new information circulated about nature, age-old ideas did not disappear. Instead, novel details and theories mingled with ancient knowledge. Scholars have noted that during the two centuries following 1500 many peoples' understandings of the physical world became less religious and more naturalistic, less reliant on semi-mythical stories and more on what was emerging as a scientific method based on experimentation and observation (cf. Greenblatt 2011). However, the proliferation of new information, its circulation buoyed by the ever-expansive reach of printed books, could not supplant ancient information. Even the most learned looked for divine explanations for phenomena that could not easily be fit into normative categories. As a result, discussion of prodigies and wonders remained common (Daston and Park 1998), and at times influenced discussions of even small creatures like insects. As the entomologist Harry B. Weiss noted in 1927, his sixteenth-century counterparts "were, for the most part, repeating what the Greeks had written" (Weiss 1927: 194).

The survival of ancient ideas about insects could be seen in paintings and manuscript illuminations when artists integrated these small animals into fine works as symbolic figures. To be sure, there were changes. Many European artists depicted butterflies on their canvasses in ways that departed from earlier artistic versions. Painters, like others, focused on the remarkable transformations of the insect during its transformation from caterpillar to cocoon to brightly colored, lighter than air flying insect. At least some of them saw the process as mirroring the transit of humans and their souls. "While in antiquity, the butterfly/soul may sometimes be a foreshadowing of death," the art historian Lucia Impelluso (2003) has written, "in the Christian imagination it became a symbol of resurrection, by virtue of the insect's journey through life and especially its emergence from the cocoon." Some painters placed butterflies into the palms of the Virgin or the infant Jesus in order to draw attention to the reality of resurrection for their god and the possibility of it for faithful believers. Artists often put butterflies in direct contrast to house flies. The butterfly signaled the rebirth of the soul and could be used, as it was by the painter Pisanello in an early fifteenth-century portrait of a recently deceased young woman, to provide comfort to the grieving with a confirmation that her soul was immortal. The fly, in contrast, almost always signified putrefaction and destruction. The Hebrew name *bá 'al zebūb* (or Belzebub in Old English), used five times in the second Book of Kings, referred to an ancient Philistine god and translated as "lord of flies." As Impelluso noted, the emergence of the name Beelzebub for the devil meant that the fly "ended up becoming a symbol of torment and torture, and therefore of the Passion of Christ." Quite a contrast to the bee, employed by artists to refer to the sweetness of Christ's relationship to humans, and the stag beetle, a surprisingly popular artistic motif intended, in the hands of northern European artists in the later years of the Renaissance, to symbolize the devil. Painters and manuscript illuminators often placed stag beetles in a menacing position on the canvas, evident in an illustrated 1526 missal in the monastery library at the Abbazia di Novacella in South Tyrol. This image depicts a stag beetle about to attack two miniature cherubs and their loyal hound. Perhaps it was the association of this insect with evil that prompted European children to tie strings around live stag beetles and taunt them to fly in circles (Impelluso 2003: 330–41, quotation at 330).

By the latter decades of the sixteenth century, what might be seen as the triumph of naturalism could be detected in accounts about insects. There were not as many treatises about these creatures as other animals. Edward Wotton (1492–1555), the first prominent English Renaissance scholar of insects, trained as a physician at Padua, one of the leading universities of Europe at the time. With patronage from the Duke of Norfolk and the Countess of Salisbury, Wotton had the time to engage his intellectual passion—the study of zoology. He published a work entitled *De Differentiis Animalium* in Paris in 1552, but never completed his natural history of insects. However, his writings about insects survived. So did those of Conrad Gesner (a.k.a. Gessner) (1516–1565), a Swiss polymath who produced a multi-volume encyclopedia about animals published from 1551 to 1587. The work totaled 4,500 illustrated pages, a remarkable achievement by any measure. But Gesner, also a physician, died of plague in Zurich in 1565, before his major work was printed. At the time of his death he had not seen his study of insects in print. It would appear over twenty years later as *Historiae Insectorum Libellus, que est Scorpione* (Gesner 1551–1587, 1587; Weiss 1927; Doran 1940). Gesner, as the historian Richard Hoffmann has observed, was "one of the last scholars to combine thorough humanist training and knowledge of texts with the naturalist's skills in observation and classification" (2005: 702).

After Gesner's death, Edward Topsell (*c.* 1572–1625), the prolific English natural historian, drew on the Swiss scholar's work for his enormous *Historie of Four-Footed Beasts,* published in London in 1607 and followed the next year with a similarly capacious assessment entitled *The History of Serpents.* By the 1610s he began his next major work of natural history, a volume he intended to call "The Fowles of Heaven." Topsell was an expert on birds—perhaps too much of an expert. He gathered information about every bird he could find, including some American birds that the English learned about only at the end of the sixteenth century. He planned to organize this book alphabetically. He began with A, but got no farther than C—though he had enumerated the existence of as many chickens and cocks as any scholar before him (Mancall 2018: 19–23).

Topsell was a compiler, not a field biologist. He drew vast amounts of his information from the works of others, especially Gesner. In an epilogue to his readers at the conclusion of his 1607 masterpiece, Topsell wrote that his god had created four categories of creatures that he needed to study: "Beasts, Birdes, Fishes, Serpents," but he added "and all creeping imperfect creatures" (Topsell 1607: sig. Yyy2^{r}). That comment might suggest that Topsell never saw insects as equal to mammals and serpents, the species that dominate his two large books. But in fact, he integrated some insects into his study of serpents: bees, wasps, hornets, caterpillars, and earthworms (Topsell 1658: 637–60, 665–71, 811–18). The timing here is crucial: Topsell, though he did not credit his source, wrote a far less exhaustive study of insects than what had already become the most important work in Europe, finished a decade before Topsell's books began to appear in print: Thomas Moffet's *Theater of Insects*.

During the sixteenth century, literate Europeans could read about the wonders of nature and human civilizations in a wide variety of books, including in the booming field of cosmographies, dazzling compendia similar to modern encyclopedias. Cosmographies, for the most part, were the creation of generalist scholars who had done copious reading and put together works that claimed to be "universal" in their reach—such as the widely translated and republished *Cosmographia Universalis* by the German scholar Sebastian Münster (1488–1552) or the massive, but never reprinted, two-volume *Cosmographia Universalis* by the French royal cosmographer André Thevet (1516–1590).

Moffet (1553–1604) was similar to those cosmographers in some ways. Like the scholars who produced those works, Moffet (also spelled Moufet, Muffet, or Moffett) was keen to show the depth of his reading. As a result, he cited other scholars frequently, including Pliny, whose ancient entomological writings remained valuable over fifteen hundred years after he wrote. Moffet finished his book in 1590, but it did not appear in print for over four decades. Only in 1634 did his work reach a wide audience as *Insectorum sive minimorum animalum theatrum*—or *The Theater of Insects or Animals*. The first English translation appeared in 1658 as *The Theater of Insects*. The title page of each edition featured a beehive on the cover surrounded by other "lesser living creatures," as the English-language image had it, including flies, scorpions, and (in the Latin edition) a caterpillar. By the time the printed work arrived, it joined one other remarkable study of insects: *De Animalubis insectus* by the Bolognese scholar Ulisse Aldrovandi (1522–1605), the author of Europe's most exhaustive study of chickens. The insect works by Moffet and Aldrovandi represented the culmination of a millennium and a half of European observations about insects. Printers, eager to show off their talents, integrated detailed renderings of insects across their pages, sometimes looking like a butterfly had landed on a page or a caterpillar had slithered across it (Figure 1.2).

FIGURE 1.2 The illustrator for the 1634 printed edition of Thomas Moffet's *Insectorum Sive Minimorum Animalium Theatrum* created these images of the swallowtail butterfly. (The Huntington Library.)

Moffet's book, which existed only in manuscript form at the end of the sixteenth century, represented the high point of European insect knowledge in the waning decades of the Renaissance. Like other scholars who published work that drew extensively on others' inquiries, Moffet frequently cited those who had observed insects in earlier generations. He included many citations to Pliny and Aristotle, as well as Gesner, Wotton, and a Cambridge physician named Thomas Penny (1532–1589). Like the ancients, Moffet understood that there were multiple reasons for writing at length about insects. First, he saw his book on insects as fitting into his generation's quest to explain the workings of the natural world in more depth than any that had been written earlier. That is, his was an intellectual project characteristic of scholars who boasted about their knowledge of ancient and new information and believed that they had new information (or better translations) to offer to the public. In this sense, Moffet's book was similar to the publication projects of the French royal surgeon and anatomist Ambroise Paré (*c.* 1510–1590) who described wonders and prodigies (Paré 1982); multiple scholars and cartographers who cataloged monstrous beings (Davies 2016); and travelers who wrote vivid accounts of parts of the world only dimly understood by Europeans (Mancall 2006). But promoting knowledge for the sake of knowledge or to enhance one's reputation were not Moffet's only goals. Like others, he also wanted to describe insects to inform others of the economic benefits that they might provide. Finally, following lines of inquiry laid down in antiquity, Moffet studied insect behavior to identify models for human action. This passion led him to write at length about bees, that most sociable insect long of interest to human observers. Of course, despite his erudition, Moffet in places wrote about creatures that we would not classify as insects—including the rhinoceros, which he believed was "of the kinde of great Beetles" (Moffet 1658: sig. Ffff3^{r}).

To Moffet, studying insects could benefit human health, including by showing ways to avoid harm. "Imitate if thou canst the Spiders curiosity," Moffet wrote, "endure the sting of the Spider Phalangium, avoid the nastiness of Lice, take a Gnat out of thy throat, sleep when Flees or Wiglice bite fiercely, keep thy trees safe from Caterpillars; drive away Weevils, Tree-worms, Vince-worms, and Timber-worms." Human bodies were frail and vulnerable. But Moffet, writing in an age when nature's wonders had divine origins, linked what he saw to questions of ultimate origins. "[W]hereof as God shews his power more in this more notable Artifice of Insects," he noted, "so his great mercy is more apparent, because there is hardly any disease of the minde or body, but a remedy may be fetcht from this store-house to cure them both." The glowworm, with its fleeting luminescence that people could not mimic, had obvious religious significance. "For who is he," Moffet queried, "that beholds the vanishing light of this, that doth not fix the eyes of his minde upon Christ the lasting, true, and the chiefest light of the world and doth not call to remembrance, that holy Spirit which doth illuminate our spirits in the most obscure darkness of our understandings?" (sig. Ffff6^{v}, 980). Humans should not worry about the outward appearance of even terrifying insects. The horse leech, for example, could be a cure for "all the melancholique and mad people" when applied by a physician to a rectal hemorrhoid (1029).

Moffet, again like his entomological predecessors dating back to Pliny, offered much about bees and their behavior. He included seven chapters on them, including details on bee names, their economic utility, the nature of their domestic and ethical systems, and various sub-categories—drones, hives, and bees who acted as thieves. The bee was "the principal" in the universe of insects. Other insects had their benefits, of course, but "the Bee doth exceed them" in terms of their use in medicines, the visual and oral pleasures they conferred, and even the "ornament of the body." The bee was so important that seemingly every major Western culture named them and many had laws to protect them (980).

These insects, he argued, should be held up as models in times of political turbulence. "Bees," he asserted, "are swayed by soverainty, not tyranny, neither do they admit of a King properly so called, by succession of by lot, but by due advice, and circumspect choice." Bees, he believed, accepted "regall authority," but they did not in the process lose their "liberty, because they still keep their Prerogative of Election." Only once "their King is once made sure to them by oath" did they "in a principal manner love him." Their larger commitment to healthy body politic had its origins at the family level, where bees had "household vertues" that taught them to store up enough honey to last from summer through winter. Close observation of hives revealed that they were immaculate. Only a very sick bee would leave excrement in the hive. They knew how to find clean water. They eschewed the killing of their enemies in their homes. They tended to the corpses of their deceased allies. They never even copulated in front of other members of the hive (892–4). Reaching into his rhetorical arsenal, Moffet praised the bravery of bees, writing that there

> is no creature under heaven so bold and adventurous as they are, insomuch, that whatsoever, whether man, or beast, or bird, or wasp shall molest them, vex and seek to destroy them, they sharply set themselves against, and according as they are able wound them with their stings.
>
> (894–6)

An underlying streak of loyalty helped explain their fortitude. Young bees grew up under the watchful and protective eye of their parents. When they matured they did whatever their monarch needed, even to the point of exhaustion and collapse (Moffet 1658: 895–6).

But Moffet realized that just as bees could be models for aspirational behavior, their colonies also at times revealed that these insects, just like humans, could succumb to more base appetites. He claimed that civil wars broke out between rival "kings" within a hive. (Here Moffet, like those who preceded him, made an error since the hive was a matriarchy and not, as early modern scholars presumed, a patriarchy.) Bees also responded to potential over-crowding with what can only be called executions of some members of the community (Moffet 1658: 932, 921, 894).

Despite such shortcomings, the remarkable qualities of bees more than demonstrated their central role among insects—indeed, among all of the animals on the planet. So, Moffet wrote, bees were the supreme proof that "God did create all creatures" for the use of humans. On one level, bees' behavior set "a pattern of Politick and Oeconomick vertues, and inform our understanding." But on another level, they possessed something that humans lacked: they were fortune tellers too. This was no mere myth. Topsell offered specific examples. Bees, he reminded his readers, swarmed in the years 208, 113, 98, and 90 BCE, foreshadowing the birth of Christ. Herodotus, Plutarch, and Julius Caesar, among others, showed how bees influenced the course of events, as Moffet explained. Virgil, in the *Aeneid*, reported that bees warned of an invasion. In one open-minded passage, Moffet acknowledged that wasps, in many ways inferior to bees, also possessed the ability to communicate to humans that calamity might strike: in 190 BCE, he wrote, "an infinite multitude of Wasps flew into the Market at Capua, and sate in the temple of Mars," where people burned them—"yet they foreshew the coming of an enemy, and the burning of the City" (Moffet 1658: 905, 927).

Among all of the insects he studied, Moffet never identified any as superior to bees. "God never created a creature less chargeable, and more profitable," he wrote. "They are bought for very little money, they will live in all places whatsoever," from mountains to forests. "The poor as well as the rich gain a great return or revenue by keeping of them," he added, "and yet need they not put more in the pot, or keep a servant the more of them." He claimed that bees could support the economy of an entire village, with their hives producing "Wax, Bee-bread, Bee-glew, Rosin, Honey-combs such as no Common weale can be without." Bees aided human economies, enhanced available *materia medica*, and proved an ongoing model for behavior. While Moffet would have been the first to acknowledge that his views reflected others' opinions, perhaps no one in the Renaissance did as much to celebrate bees. Given the circulation of his ideas, from his original manuscript to the publication of the Latin edition in 1634 and then the English translation in 1658, it is safe to say that Moffet, perhaps along with Aldrovandi, did as much as anyone to circulate insect knowledge in his era (Neri 2011: 27–73). Moffet, for his part, never knew of the printed editions of his masterwork. He died in 1604, six years after he finished his work but a generation before its first printing. His work became one of the cornerstones of English entomology, though naturalists including Francis Willughby (1635–1672) and John Ray (1627–1705) made substantial advancements in the field in the latter decades of the seventeenth century (Raven 1947; Ogilvie 2012).

AMERICAN INSECTS AND ENTOMOLOGICAL KNOWLEDGE

Scholars like Moffet and Aldrovandi filled their learned books with knowledge long available in Europe. But they also wrote at a time when Europeans had been wrestling with the growing body of information about American nature since the first voyage of Christopher Columbus over a century before Moffet completed his work. One travel account after another added to Europeans' understanding of the marvels to be found in the Western Hemisphere. Together, they expanded Europeans' sense of the catalog of nature and represented ongoing proof of the ways that knowledge crossing the Atlantic altered Old

World sensibilities (Elliott 1970; Grafton 1992; Greenblatt 1992; Lowood 1995). So did herbalists, who integrated American plants and insects into their works, especially insects that were unlike any known in Europe. Here they could draw on accounts like the "Histoire Naturelle des Indes," which offered an illustration of a firefly. The unknown author of the caption wrote that this insect "has such power as to give a light at night so strong that one can easily recognize people from a distance of two or three paces, and it flies in the air" (*Drake Manuscript*, f. 78^{r}, p. 263).

Among those who shared views about American nature was a Huguenot missionary named Jean de Léry (1536–1613), who sailed from France to Brazil in 1556 to help establish a community in territory previously claimed by the Portuguese. Two years later he returned to France. Unlike other travelers, who often rushed to get accounts of their exploits into print, the first edition of Léry's report appeared in print in Geneva in 1578. (While the official travel account took twenty years to appear, Léry did earlier write a book about Brazil, which was published in 1561.) Like some—but not all—European accounts of the Western Hemisphere, Léry's was sufficiently popular to appear in five more French-language editions by 1611. By then, editions had appeared in Latin, including a now-famous illustrated account by the Frankfurt-am-Main shop of the Flemish engraver Theodor de Bry (1528–1598), as well as in German and English versions (Léry 1990: xvi–xviii, 257–9).

Léry's report contained detailed information about the voyage across the Atlantic and the ways that sailors and other travelers understood the marine creatures they encountered, including sharks and flying fish. He also turned his attention to creatures in Brazil, including insects. The first to attract his attention was, not surprisingly, the bee. Brazilian bees were not like the European variety. Instead, he wrote, they "rather resemble those little black flies that we have in the summer, especially in the grape season." Like Old World bees, they produced honey and wax, though they did so "in the hollows of trees, where the savages knew how to collect them." The Tupinambás, he noted, combined the honey with the wax to make what they called *yra-yetic*. But they also separated the two substances. They ate the honey, like Europeans. But they took the wax, which was black, and rolled it into cylinders the size of a man's arm. Curiously to him, the Tupinambás did not use them as candles, for which they had no need since they employed local tree branches to light their paths. Instead, they used the wax to caulk "big wooden tubes where they keep their featherwork, to preserve them against a certain kind of moth which otherwise would spoil them" (92).

Bees were not the only American insect to attract Léry's attention. He also wrote about cricket-sized insects that the Tupinambás referred to as *aravers*, which modern scholars believe encompassed a range of insects including scarabs, beetles, and cockroaches. These voracious creatures swarmed near fires in the night seeking food. They gnawed leather collars and shoes. They preferred chickens, which they "would gnaw down to the bone," Léry wrote, "and we could expect to find skeletons in the morning" (92–3, 240 n. 13).

Léry wrote at greater length about other insects. Europeans suffered from a Brazilian gnat-like creature the Tupinambás called *yetin*. It had a potent stinger that could penetrate the visitors' clothing so precisely "that you would think it was a needle point." He wrote about the scorpion, a creature he included along with insects. These were dangerous animals, as he learned from experience when one stung him. Léry quickly went to find medical help, which came in the form of an oil made from crushed scorpions, the same treatment that the Tupinambás employed. The Frenchman did not note if the Europeans learned about this cure from their hosts. He saw the cure in medical terms. The Tupinambás, by contrast, had developed the cure as a way to punish the offending animal by murdering it and using its remains as medicine. Whatever the origins of the treatment, Léry suffered for a full day and, as he wrote, could hardly handle "the violence of the pain" (94).

Léry's most extensive entomological observations focused on a flea-sized tormenter the Tupinambás called a *ton*. Modern entomologists have identified this as *Tunga penetrans*, the chigoe flea, also known as a jigger. His description of it revealed a danger as great as the mosquitoes that swarmed the artist of the "Histoire Naturelle des Indes." This *ton* dug under a person's nails, producing an almost insatiable itch. Intelligent people removed the offender as quickly as possible because if left alone it would dig deeper and produce a welt the size of a pea "so that you cannot pull it out without great pain." As might be expected, the often bare-footed Tupinambás suffered the *ton*'s attack, but Léry was surprised to learn that it burrowed its way into the feet of well-shod French travelers too. He knew what to look for, but could not prevent their incessant attack. He warned his readers that he suffered at least twenty bites each day. Those who paid less attention suffered from painful sores on their hands and feet and also on what he called "other softer parts of the body," including the armpit. He reported that the Spanish natural historian Francisco López de Gómara had also described them and gave the same advice that Léry offered to readers: the *ton* would produce no harm if victims or their friends removed them quickly (93).

Had he stopped there, Léry's book would have been like many others produced by Europeans: an observation of an insect, its impact on humans, and ways to render the dangers it posed less menacing. But Léry added more information to his book. Specifically, he wrote about how the Tupinambás handled the *ton*. "To get rid of them," he wrote, "our Americans rub the ends of their toes or other parts of the body where they try to nest with a thick, reddish oil, made from a fruit they call *couroq*," which modern scholars have identified as nutmeg (*Myristica officinalis*). Léry compared the *couroq* to a European chestnut, using the well-worn strategy of describing something unfamiliar to readers by comparing it to something they would know. As it turned out, the oil produced by the *couroq* could be used in a variety of cures, including as a salve for wounds or broken bones. The Tupinambás, he noted, "knowing its virtue, consider it as precious as some people over here regard what they call holy oil." The ship's surgeon who accompanied Léry homeward brought back large jars to conduct experiments (93).

Léry shared his understanding of American insects in his travel account, a genre that had become very popular across Europe during the sixteenth century. Information about the insects of the Western Hemisphere spread in other ways as well. Most notable was information that moved across the Atlantic in the voice of indigenous Americans, or at least their proxies: European scribes who learned Native languages and wrote down what they heard, as well as Americans who learned to write formerly non-written languages and to share their information with images as well as words. Perhaps the most thorough of these accounts for the period before 1600 was a book by a Spanish Franciscan, Bernardino de Sahagún (1499–1590), who worked with Mexica (known to the Europeans as Aztecs) scholars to create a vast, richly detailed codex that he called the *Historia general de las cosas de la Nueva España*, or the *General History of the Things of New Spain*. It is also known today as the *Florentine Codex*. The work includes text in Nahuatl, the language of the Mexica, Spanish, and a large number of small illustrations, many (though not all) in color.

Sahagún's informants had much to tell about the natural world, including about insects. They told how the *çacatecuilichtli*, a caterpillar, built its cocoons only in cherry trees; this was better for the insects than for the tree, which no longer bore any fruit once it had been colonized. Like European observers, those who knew about American insects made value judgments. They called the *tlalchapolin*, also known as *ixpopoyochapolin*, a "stupid" creature because females delivered their young in the middle of a road and did not move even when they might be crushed. The informants reported that the *tapaxi* (or *tapayaxi*), which lived in the ground, was notable for the performative aspects of its procreative acts. Another ground dweller was the *tecuitlaololo*, a creature that spent its time making balls out of dung and rolling them around. According to the cleric, the insect seemed to have no other reason for its existence (Sahagún [1540–1585] 1950–1982, 11: 96–7, 92–3).

Like sixteenth-century European entomological writers, the Mexica informants paid close attention to the social structure and observable behavior of small creatures. There were multiple species of ants in New Spain, and the *Florentine Codex* contains descriptions of the behavior of some. The text noted that there was a red ant called *tzicatana*, known as "the conqueror" for its practice of moving in large swarms and for its voracious appetite. Locals reported that "they draw up for battle." Others were less violent. One was

FIGURE 1.3 This image of two men harvesting cochineal, labeled "The nopal plant that is grown in America and produces grana" (or insect dye) appeared in a report on Catholic dioceses in New Spain and Peru (1620–1649). Reprinted from the Newberry Library, Chicago (open access) (Vault Ayer MS 1106 D8, Box 1, folder 15).

known for its domestic habits: it "builds a nest for itself, hunts its food. It is a carrier, a transporter, a storer of things. It is a wood-dragger, a dragger of things." Most ants moved in packs, including the *quauhazacatl* or tree ant and the *cujtlaazcatle* or dung ant. Yet one, the *icel azcatl*, known for its bright red exterior, traveled alone (Sahagún [1540–1585] 1950–1982, 11: 89–91).

Much of the information about Mexican insects available in this codex detailed behavior and appearance. But this text, again like others by Europeans, emphasized the utility of specific insects for humans. Sahagún's informants provided details about the *conyayaual*, a centipede that tended to make itself into a ball when threatened. But what mattered more was the way that the Mexica used it. As the codex reported, they gathered up these centipedes and crushed them to make a salve that soothed pain ranging from gout to a toothache. Hard-won experience of this insect taught a vital lesson: used in moderation it worked as a balm, but if someone consumed too much he or she could die (11: 92).

Two insects drew extensive discussion: four species of fireflies and a scale insect known as cochineal. The luminescence of the *icpitl* set them apart from the millions of other insects crawling, burrowing, and flying around New Spain. The text described one known as the *coquitl* as looking like a locust, but with a light-emitting section that took up half of its length. It tended to come out at night after a rainfall. "Sometimes it shines like a pine torch, because of the dark," the text's scribe wrote. At times the *coquitl* seemed to travel in a line. This action prompted some observers to call them *tlauipochtin*, the term for sorcerer.

More important by far was the cochineal, a soft-bodied sedentary scale insect found on a prickly pear cactus tree called a nopal. Pregnant women who feared a breach birth ate cochineal when the insects were alive, believing that the animals would cause the fetus to move in the womb. That knowledge might have come in handy for obstetric practice in Europe, but rather than being known for its potential health benefits when consumed alive, the cochineal instead became known for its color. As it approached the end of its life, the insect became engorged with fat. Those who walked by a nopal might have thought there were blood-colored blisters on its trunk and branches. Locals harvested these insects (see Figure 1.3), dried them, and crushed them. The result was a vivid red dye that was good quality and less expensive than the Polish cochineal and the Armenian cochineal already being used by Europeans. The news spread quickly. One natural historian after another wrote about the wondrous dye to be made from these American insects. "[T]he famous cochineal of the Indies," the naturalist José de Acosta (1540–1600) wrote in his account of the Americas, produced "the best red dye." Merchants arranged for the shipment of the dried insects (grana) to Spain in "a rich and voluminous trade" (Sahagún [1540–1585] 1950–1982: 100–1, 122, 180, 239–40; Acosta 2002: 213; Phipps [2010] 2012).

THE RENAISSANCE FOR INSECTS

Two major developments stand out as the most significant in the era of the Renaissance: Gutenberg's invention of movable type and the meeting of Europeans with Americans. The first development facilitated the spread of information about the natural world, including about insects. Details about American insects could be found not only in travel accounts like Léry's but in herbals too, compendia of pharmacological information primarily based on finding medicine through plants and also, on many occasions, from insects. The second development benefited Europeans, who found what was to them an unknown hemisphere and set about to harvest its natural resources, a process that began with Columbus himself. But extracting biota from the Americas, including insects like cochineal, produced devastation for indigenous Americans when they came into contact with Europeans or their animals bearing disease-causing viruses or bacteria. Well before 1600, the process that the environmental historian Alfred Crosby labeled the "Columbian Exchange" had enriched many Europeans and killed countless Americans. Their deaths did not result from planned genocidal campaigns, though there were ample moments of astonishing violence. Instead, the process of European conquest and colonization placed new stress on American bodies and made Indigenous men, women, and children more likely to succumb to invasive pathogens (Crosby 1986; Jones 2003).

The cultural movement known as the Renaissance had kindled in early modern Europeans a thirst for ancient knowledge and a celebration of scholars like Pliny, whose works found new audiences, thanks in part to the spread of the printing press. In the hands of proto-entomologists like Thomas Moffet, antique and modern information about the smallest animals came together, providing readers with a more exhaustive and reliable guide to the universe of insects. The arrival in the continent of new biota from the Americas, including useful insects like cochineal, had economic benefits, which emerged at the same time as European travelers warned the readers of their narratives about the dangers posed by insects like the mosquito and the fire ant. Armed with observations based often on Indigenous knowledge about how to fend off the worst aspects of living with dangerous insects, information about these small animals encouraged the expansion of European claims.

In 1300 it was common for many Europeans, at least, to look at an insect in a painting and see it as a metaphor. Three hundred years later, the vast majority of illustrations about insects could be found in works of science. Natural historians and printers created an intellectual agenda that helped shape the reports of those who went on long journeys. Scores of travelers offered accounts of places far from their homes, often including observations of insects in their published work. This desire to describe unknown insects continued for generations. The German naturalist Engelbert Kaempfer, a physician working for the Dutch East India Company who arrived in Japan 1690, wrote a widely read account of its culture and environment, including its insects (Kaempfer 1999: 72–3). In the years after 1600, information about these small creatures spread more quickly than ever before, offering diverse communities information about how to prosper in a world of insects.

CHAPTER TWO

Insects and Disease

Unsuspected Agents of Death

MICHELLE ZIEGLER AND SHERILYN G.F. SMITH

Insects are ubiquitous, and living in intimate contact with insects can be assumed to be a fact of life for everyone during the Renaissance. Such contact had consequences for human health on two levels, one direct—a consequence of bites or infestation—and the other the result of disease transmission by insect vectors, generally through the bites of blood-feeding insects. Both had impacts on human populations and culture during the Renaissance, the former being recognized but perceived as minor, a nuisance, or in serious cases, a punishment from God. In fact, the Bible describes lice as one of the ten plagues on the Egyptians (Exodus Chapter 8), and so lice may well have been regarded as a divine punishment for sin. The latter impact, disease transmission, was unrecognized but had devastating effects. The most widely recognized of these diseases was the Black Death, which ravaged Europe in the mid-fourteenth century.

Understanding the impact of insects on human health requires an understanding of the physical and social conditions of the time, historical disease conditions such as areas where diseases were already endemic, and the state of medical knowledge. It also requires sources of information such as written descriptions, medical records, and other evidence that can tie insects, either directly or indirectly, to disease. That is particularly challenging during the Renaissance; the germ theory of disease had not yet been formulated, and the physicians of the time attributed disease to such things as unfavorable conjunctions of the planets, or foul-smelling miasmas emanating from swamps, decomposing organic material, and volcanoes. Foul odors were seen as the cause of the generation of imbalances of "humors" in the Classical sense and therefore disease. Even descriptions of disease are ambiguous; the term pestilence is used for any outbreak of disease (and could therefore be any of a number of actual diseases), and so causal agents and means of transmission can be challenging to identify.

Physical conditions, such as topography, climate, and seasonality, are important determinants of insect presence and population size. Climate can influence the relationship between insect vectors (transmitters of disease) and outbreaks of disease in humans or reservoir hosts (Campbell 2016). Insect development though all life stages—egg, nymph, or larva, pupa, and adult—is temperature-dependent, and temperature patterns also can alter the length of the pupal (cocoon) stage in different seasons. For example, *Pulex irritans*, the human flea (and a vector of bubonic plague), goes into an extended pupal stage in late fall, resulting in no adult emergence in the winter. The surge of fleas in the early spring is likely to be caused by the emergence of adults from all of the pupae whose development was arrested by colder temperatures in the fall and early winter (Christodoulopoulos et al. 2006; Ziegler 2016b). Even so, by living in human homes and livestock shelters, *P. irritans* has managed to survive in arctic conditions as far north as the western shore of Greenland and the Orkneys, and fleas and lice have been found throughout Europe as far north as the Arctic circle (Panagiotakopulu and Buckland 2017). These insects, which can be a vector for a variety of diseases as well as being able to infest humans and animals, would have survived and even thrived in the cool wet conditions of the Little Ice Age that pervaded throughout Europe during the Renaissance (Reiter 2000). Conditions of the Little Ice Age led to crop failures and famine, which in turn led to undernourishment and a higher susceptibility to disease, including plague and other insect-vectored diseases (Appleby 1980; Stenseth et al. 2006; Schmid et al. 2015; Campbell 2016).

Similarly, in the early modern period, malaria transmission by *Anopheles* mosquitoes wintering inside homes was recorded north of the Arctic circle (Reiter 2000, 2001). The presence of malaria in England and elsewhere across Europe during the Renaissance has been widely documented (Franklin 1983; Dobson 1989; Reiter 2000; Schats 2015; Newfield 2017; and others). Another important mosquito species is *Aedes aegypti*, a vector of yellow fever and a biting pest that had become adapted to thrive within and around domestic spaces. Its invasion of the Americas during the sixteenth century, accompanying European colonization and the beginnings of the slave trade, resulted in the spread of yellow fever, sometimes called the American plague, which would become widespread in the seventeenth century. The ability of all of these insects to survive and maintain population levels that could impact human health is tied to ever-changing climatic conditions, and disease outbreaks frequently connect to those conditions favoring insect vectors, pests, and/or parasites (Stenseth et al. 2006; Schmid et al. 2015; and others).

Living conditions also contribute to the exposure of people to particular types of insects and insect vectors (Mitchell 2015). Both rural and urban populations were likely to encounter insects associated with farm animals, such as various biting flies and nuisance flies, all of which could potentially transmit disease. Proximity to bodies of water and low-lying marshlands, or those utilizing such areas to make a living such as watermen, would have increased exposure to a different set of potential disease vectors. Urban populations, crowded into smaller geographic areas and exposed to variable sanitary conditions and water supplies, as well as air quality, might also be expected to fall victim to diseases transmitted directly or through the air (Rawcliffe 2012, 2013). At the same time, populations of some insects such as houseflies and cockroaches might be expected to increase in a more stable urban environment with more plentiful food supplies. The living conditions of the wealthy also certainly differed from the living conditions of peasants or the emerging middle class (see Kelly and O'Grada 2014). Therefore, the types and prevalence of insect-borne and insect-caused diseases should vary depending on all of these factors. Assessing the impact of insects on human health during a time when the causes and mode of transmission of many diseases remained unknown offers unique challenges.

To uncover the impact of insects on health and disease in the past, scholars today must creatively search for evidence, especially evidence that confirms or corroborates written historical records, which can sometimes be ambiguous. One of the best sources of evidence for the direct impact of insect pests such as biting insects is the discovery of archaeological remains of insects in areas of historical interest or significance (and that are unlikely to have been invaded more recently, such as a flea found within a lead-sealed coffin). Modern paleoarchaeologists are currently searching new excavations as well as older sites such as graves to find specimens or parts of insects that might suggest something of the importance of those insects to health; it is possible to identify even very small specimens, such as flea eggs. Sites being searched include church crypts and mausoleums for the oldest sealed caskets as well as graves and their surrounding sediments to look for insect remains. Reliquaries are another possible source; indeed, a parasitic flatworm called a schistosome has been identified inside a reliquary containing some of the abdominal organs of St. Louis, King of France (1214–1270) (Charlier et al. 2015). Schistosomes are soft-bodied, and so this discovery opens the possibility of finding insects or insect fragments in such sites.

Finding tools used specifically to remove an insect, like the louse (or nit) comb, is another type of evidence, in this case for the presence of head lice. A louse comb is usually a relatively short comb with fine teeth on one side and ultrafine teeth on the other side to remove eggs (nits). They were often made of ivory or bone for durability. Such combs have been found dating to the fourteenth through sixteenth centuries. Woodcuts in *Hortus Sanitatis* (*c.* 1491) show infested bedclothes and the removal of lice (see Figure 2.1).

The most recent technology that can reveal the presence of insects and/or the diseases they may have transmitted is the extraction and sequencing of ancient DNA, termed aDNA, from human (or other) remains. Any pathogenic organism that causes a systemic infection will be widespread in the victim. While soft tissues decay away, teeth usually remain intact, and pathogens within the tooth pulp leave a tell-tale signature in the form of DNA that can be isolated and sequenced. In ancient human remains, this can be a delicate process, as it is often a challenge to extract enough DNA for sequencing and to correct for degradation of the DNA molecule over time in the interpretation of the results. However, properly sequenced DNA is a powerful piece of evidence: if an insect-transmitted disease is present, then the insect itself must have been present as well. In the last two decades, DNA technology has opened up a new avenue to discovering evidence of the

CAPVT. 118.

PVlex Ex libro de naturis rer. Pulices uocati ſūt, eo q̄ in puluere magis nutriūt. Patet pulicem eſſe uermiculū nigrū & minutum quidem, ſed ualde pūgitiuum, maxime autem tempore ęſtiuo & pluuiali. Saliunt autem potius q̄ uolāt, nocte q̄ die magis hominem infeſtant. Et niſi uehemens corpor aut ſomnus grauis in membris ſit, ſe tueri non poteſt. Cum quis manum ad-

luto ſepe infuſo.

CAPVT. 119.

PEdiculꝰ. Iſi. Pediculi ſunt uermes cutis: à pedibus dicti: unde pediculoſi dicti ſuut, qbus pediculi in corpore efferueſcunt. Ex libro de naturis rerum. Pediculi dicuntur à numeroſitate pedum: hoc malum ex ipſa hominis carne creatur indubitanter, & tamen inuiſibiliter, hos nonnul

FIGURE 2.1 Left: Fleas or lice being shaken from the linens; Right: a woman brushes lice from a boy's head. *Hortus Sanitatis*, *c.* 1491, courtesy of the Wellcome Collection, CC BY 4.0.

presence of insects and, perhaps more significantly, to discovering molecular evidence of insect-borne (and other) diseases in human remains. Raoult et al. (2000) used PCR to identify *Yersinia pestis* DNA in archaeological remains from the Black Death in the fourteenth century. Subsequently, molecular techniques have enabled the identification of *Y. pestis* elsewhere, as well as louse-borne epidemic typhus, trench fever, relapsing fever, malaria, visceral and mucocutaneous leishmaniasis, and a variety of bacterial and viral diseases which in some cases were co-infections with an insect-borne disease. Ancient DNA from archaeological remains, along with immunological evidence that can now be used to indicate infection, have become major tools for understanding the distribution, spread, and ecology of insect-vectored diseases in the past.

Words, of course, provide evidence as well: records, references to, or descriptions of an insect-transmitted disease, or a disease directly caused by an insect infestation, or texts giving remedies for such a disease are invaluable evidence. At the same time, they must be interpreted carefully because of the lack of knowledge about disease or its cause during the period. For example, malaria, then known as tertian or quartan fevers or ague, has been commonly reported in Europe since at least Late Antiquity and through the twentieth century. "The topographer, John Norden, visited Essex in the 1590s but was unable to 'comende the healthfulness of it: And especiallie nere the sea coastes ... and other lowe places about the creeks which gave me a most cruell quarterne fever'" (Dobson 1989: 3). Reiter (2000) also describes malaria as causing mortality in England in the sixteenth century, and Dobson (1989: 3) states:

> As early as the 16th century the English marshlands were singled out for their extreme unhealthiness. Residents and visitors described the "marsh fevers" of these districts, variously calling them "agues", "tertian fevers", "quartan fevers" or "intermittent fevers" and attributing them to the noxious vapours of the stagnant marshes.

Similar conditions must have occurred elsewhere in Europe. (Most European countries did not eradicate malaria until after the Second World War.) The species of *Plasmodium* that cause malaria in humans can only be effectively transmitted by mosquitoes in the genus *Anopheles*. In these cases, if the disease is determined to have been present, then so must have been its insect vector.

Remedies are also known from collections explaining how to treat various conditions, including infestation with biting insects. For example, a remedy for lice infestations from "a circa 1600 compilation of medical remedies, is a drink consisting of cheese whey and a little vinegar. Imbibed on 'certain days,' it will cause all the lice to die, and 'there will breed no more about you'" (Folger MS

V.a.140, f. 4r, qtd. in Wolfe 2018). Another remedy from the same collection involves using powdered antler from the red deer, making both a drink of it and rubbing it on the head (Folger MS V.a.140, f. 3r, qtd. in Wolfe 2018):

> To kill lise, and nittes in the heade.
> Take the pouder or scrapinge of Hartes horne, &
> make the pacient to Drinke of it, and ther will no lise
> nor nittes breede in his head, but if yow strowe the sayde
> pouder vpon his heade, all *th*e lise & nittes will die.

Again, if a remedy for a condition is known, the condition can be assumed to have existed, even if not correctly diagnosed.

INSECT INFESTATIONS: LICE, FLEAS, AND BEDBUGS

In Renaissance Europe, infestation with lice (as well as fleas) was not uncommon, as depicted in paintings and woodcuts such as the 1491 woodcuts of linens infested with fleas or lice and a woman brushing lice from the head of a boy (Figure 2.1). Louse (and perhaps flea) infestations were treated with topical medications containing mercury compounds (quicksilver), as well as extracts of *Delphinium staphisagria* (lice-bane) and arsenicals (Fornaciari et al. 2011; Vicentini et al. 2018), in addition to the more easily obtainable powdered deer antler. Such treatments show that infestations were considered conditions requiring medication. Nonetheless, such infestations were also seen to encourage religious contemplation on vanity, or were seen as a penance (Smith 2019). Many Christian saints were praised for never bathing, and enduring such infestations.

If in some medieval and Renaissance reports of infestations, lice, fleas, and other ectoparasites sometimes seem mixed up, it is because they were. Lice and fleas, like rats and mice, were often not clearly distinguished by seventeenth-century writers (Cole 2016: 17), and the same lack of distinction may have occurred during the Renaissance. Issues of translation may also have contributed to confusion; even in the Bible, passages including insects differ in the insect named in various translations. Kritsky and Cherry (2000) note that the lice that were inflicted on the Egyptians as the third plague, as described in the King James version of the Bible, were probably maggots, based on translation of the original texts. Descriptions of fleas, lice, and other "vermin" were often translated from older writings in Latin, Greek, and other languages (Aristotle's *Historia Animalium* is a prime example), and so some confusion is understandable. At the same time, writings such as the remedies above and poetry exist that clearly deal with a specific parasitic insect:

Marke but this flea, and marke in this,
How little that which thou deny'st me is;
It sucked from me first, and now sucks thee,
And in this flea, our two bloods mingled bee;
Thou know'st that this cannot be said
a sinne, nor shame not loss of maidenhead,
Yet this enjoyes before it wooe,
And pamper'd swells with one blood made of two;
And this, alas, is more than wee would doe.

John Donne, *The Flea* (1633)[1]

In the late Renaissance and through the seventeenth century, fleas inexplicably became objects of envy and were imagined to have quite discriminating taste in their choice of hosts. Fleas gained the respect of agency, control over their own actions (Cole 2016: 7–8). Young John Donne (1572–1631), later to become the famous church cleric and poet, clearly envies the flea that wanders the flesh of his beloved. Donne could hardly have written a better description of a disease vector (transmitter of disease) if he had had a perfect understanding of insect-transmitted disease, but he never suspected the role of fleas in disease transmission, nor did anyone else in the Renaissance. This was only the beginning of a romanticized fascination with fleas; artists such as La Tour, Crespi, and van Honthorst painted erotic scenes of young women searching for fleas, and such paintings of "flea hunting" became a genre of seventeenth–eighteenth-century painting (Potter 2013; Cole 2016: 7–12).

Romanticized notions aside, fleas of several species are associated with humans and domestic animals, particularly *Pulex irritans* (the human flea), species in the genus *Ctenocephalides* (which includes dog and cat fleas), and rodent fleas, particularly *Xenopsylla cheopis* (the oriental rat flea). Fleas have been recovered at archaeological sites including Dutch sites dating from the eleventh to the fourteenth century (Schelvis 1994), medieval Dublin, and from the western Viking settlement in Greenland (Buckland 1988; Sadler 1990; Forbes et al. 2013). Busvine (1976) describes fifteenth-century women's stoles of animal fur, worn around the shoulders and known as "flea furs," while "flea traps" made of small animal skins were also worn like a necklace. The biting capacity of fleas with the accompanying inflammation and itching of bites was clearly recognized. The ability of fleas to transmit diseases such as plague (caused by *Yersinia pestis*) and epidemic typhus (caused by *Rickettsia prowazekii)*, as well as serving as intermediate (infective) hosts for the tapeworm *Dipylidium caninum*, was not understood during the Renaissance, but clearly these insect vectors had a monumental impact on Renaissance society.

Lice are one of the few insects that have been known to infest humans for almost as long as we have been human (Panagiotakopulu and Buckland 2017;

Panagiotakopulu et al. 2018). Humans can be infested by three kinds of lice: head lice (*Pediculus humanus capitis*), body lice (*Pediculus humanus humanus*), and pubic lice (*Phthiris pubis*). All three varieties of human lice were found in all economic classes and at all stations of life during the Renaissance. Eggs (nits) are laid on hair or occasionally clothing, and the adults can live up to three months, feeding on human blood. Lice fare better at temperatures above freezing with a relative humidity of 70–90 percent. The close association of lice with the human body provides enough heat and humidity for them to be found in all conditions suitable for humans, from tropical to near arctic conditions. Lice cannot, however, tolerate temperatures over 50 degrees Celsius (122 degrees Fahrenheit) or humidity below 40 percent (Badiaga and Brouqui 2012: 333).

Evidence of lice infestation has been found in mummies in church crypts, such as the crypt of Ferdinand II of Aragon (1467–1496), King of Naples (Fornaciari et al. 2009). Sealed in an inner lead coffin in the Basilica of San Domenico Maggiore in Naples, approximately half of his body had remained intact, with the rest having been skeletonized. The young king had only been on the throne for slightly over a year, placed there by his father Alfonso II of Naples specifically to lead the army against King Charles VIII of France. He succeeded in defeating the French and reuniting all of the kingdom of Naples. He then married his father's half-sister, Giovanna of Aragon, but only a few months later he died of malignant tertian malaria, on October 7, 1496 at the age of twenty-nine (Fornaciari et al. 2011: 110).

Archaeologists were able to examine Ferdinand II's hair and skin for mercury, a toxic element in ointments commonly applied to the skin to treat lice. Microscopic examination revealed many fragments of lice in the hair of the scalp and pubis. Ferdinand II's louse fragments were too small to distinguish the species, but given their very specific localities on the body, they are assumed to be head and pubic lice. Researchers also found a very high level of mercury in the hair and skin of the scalp, but not in the pubis, skin elsewhere, or internal remains. The source of the mercury was most likely cinnabar, a widely traded mineral compound of mercury sulfide (HgS), typically used as a red pigment, which can be distilled into a concentrate of pure, liquid, elemental mercury, also called quicksilver. The extremely high levels in the hair and scalp alone can only be explained by external application as a wash. Ferdinand II's remains are the first to show signs of the use of external mercury to cope with a parasite. These archaeological findings also illustrate that fifteenth-century kings were not immune to these irritating hitchhikers (Fornaciari et al. 2009, 2011).

Human fleas and lice came to be taken for granted, so much so that in the Ottoman Empire, lacking them could be seen as a sign of disease. Ottoman explorer Evliya Çelebi (1611–1682), after observing infested people throughout his travels, wrote that "the louse is of delicate nature, it likes clean places, does not like the leper body" (Varlik 2015: 34). For the Ottomans then, a lack of fleas and lice indicated that person was a leper or had a foul body odor. They took this so seriously that the marriage of Rüstem Pasha to Mihrimah, a daughter of Emperor Süleyman the Magnificent (1494–1566) was nearly derailed because the Emperor heard that he lacked lice. A doctor was dispatched and he managed to find a single louse, saving the marriage. This story circulated among the Ottoman elites as an example of the good fortune of the grand vizir Rüstem Pasha; even his lice brought him luck (Varlik 2015: 34–5).

The immense Ottoman Empire, with territory stretching around the rim of the Mediterranean from Greece to Egypt, played an integral role in the disease landscape of the Renaissance. European travelers to the East often mentioned the common secondary clothing markets, called *bit pazari* (flea markets), teeming with fleas and lice that contemporaries identified as a source of plague. A relationship between flea markets and plague, but not fleas and plague, was recognized by the mid-Renaissance. (Yet even in the nineteenth century, French physician Francois Pouqueville (d. 1838) wondered at how much *miasma*, "contagious" foul air, was produced by the used clothing at these markets.) Filth and vermin were thought to be the source; fleas and lice were still not recognized as agents transmitting disease. Indeed, lice were thought to arise by spontaneous generation from filth.

The bedbug (*Cimex lectularius*) is the third insect species associated with infestation. Bedbugs have been human companions since ancient times, and archaeological evidence of bedbugs has been found in medieval England (Kenward and Allison 1994), on combs in fourteenth-century sites in the Netherlands (Schelvis 1994), and in the Americas following colonization in the fifteenth and sixteenth centuries (Bain 2004). Bedbugs live in bedding, rugs, and other domestic spaces, and come out at night to find a host and feed on blood, causing inflammation at the site of bites (although because they have an anaesthetic in their saliva, the biting itself is not felt). Historically they have not been implicated in the transmission of disease, although recent work has shown that bedbugs can become infected with over forty pathogens (Delaunay et al. 2011) and can transmit trench fever, caused by *Bartonella quintana* (Leulmi et al. 2015) and *Borrellia recurrentis* which causes relapsing fever (El Hamzaoui et al. 2019). Plague, tuberculosis, and Chagas disease are among those pathogens that can infect bedbugs, but the competence of bedbugs to transmit these diseases is unknown (Cholewiński, Derda, and Hadaś 2017: 85). Because these diseases were not recognized as distinct from other diseases in the fourteenth through sixteenth centuries, there are no accounts linking disease to bedbugs, if in fact such links existed. There are, however, seventeenth-century suggestions on how to get rid of bedbugs, including using the smoke of ox dung, lupines, arsenic, cypress, and horsehair (Busvine 1976: 190), suggesting that bedbug infestations were not uncommon.

INSECT-VECTORED DISEASES: PLAGUE, RATS, AND FLEAS

The dark specter over the Renaissance is the pandemic known as the Black Death. Both it and recurrences of plague ("peste" or "pestilence") throughout the remainder of the Renaissance (and through the eighteenth century, the "Second Pandemic") have been the focus of considerable historical and biological attention. Numerous written records exist, including manorial accounts, descriptions from monasteries and dioceses, mortmain (death tax) records, and chronicles (Horrox 1994; Dobson 1997; Aberth 2005; and others). Records of the spread of the Black Death (and subsequent recurrences) have been examined in light of distributions and movements of rodents, microclimatic factors, the presence of plague in the absence of rats in Iceland (Karlsson 1996), and other confounding factors.[2]

The advent of DNA technology has led to identification of disease agents in archaeological sites containing human remains. Ancient *Yersinia pestis* DNA (aDNA) was found in multiple sites in Europe prior to 2010 (Garrelt and Wiechmann 2003 and others), although some of those studies were controversial. Then in 2010, Haensch et al. published results of an extensive study of aDNA from putative plague victims across Europe, and that study persuaded many that *Y. pestis* was the causal agent. The extraction of the complete genome of *Yersinia pestis* from the East Smithfield cemetery in London, which was established in 1348–1349 specifically as an overflow cemetery for this specific epidemic, definitively confirmed the cause and ended reasonable debate over the etiology of the Black Death (Bos et al. 2011). This has been further confirmed by the discovery of *Y. pestis* aDNA in plague cemeteries and plague pits across Europe (Bos et al. 2011; Seifert et al. 2013, 2106; and others). The reconstruction of the *Yersinia pestis* phylogeny has made the study of these epidemics one of the most vibrant areas of research in the history of medicine, and indeed has been of tremendous benefit to our understanding of modern plague.

The Black Death reached Europe from its origin in Asia in 1346. Horrox (1994: 9) describes the movement of the plague, perhaps spreading south to China and India[3] and west to Persia and Russia, reaching that country by 1345–1346. It moved overland to the Crimea; Slavin (2019) gives a thorough discussion of outbreaks recorded from the Chu Valley in what is now Kyrgyzstan. Plague then spread to Italian trading centers on the Black Sea, and, by Genoese traders fleeing the plague, it arrived in Constantinople and in Italy in 1347. A contemporary account of the arrival of plague in Italy was written by Gabriele de Mussis, a lawyer in Piacenza (see the original translated document in Horrox 1994: 14–26). In summary, Genoese traders, fleeing attack by the Tartars who themselves were being ravaged by plague, went to Caffa, an Italian trade center on the Black Sea. The Tartars besieged the city, but plague killed so many of them that they ended the siege; they reportedly used catapults to send corpses over the walls into the city, with the intent of spreading plague. The Genoese then fled the city but carried the plague with them. De Mussis wrote, “Moreover, one infected man could carry the poison to others, and infect people and places with the disease by look alone. No one knew, or could discover, a means of defense” (Horrox 1994: 17). Wheelis (2002) accepts this description of biological warfare but argues that the incident was unrelated to the plague reaching Italy, while Benedictow (2004: 52–3) has argued that this purported use of corpses as biological weapons is “untenable,” based on the necessity of infective fleas being on those corpses.

Nonetheless, by 1348 plague had reached Bavaria, Germany, Austria, France, the Low Countries, and had crossed over to England. Descriptions, often with commentary concerning sinful behaviors (indecent clothing, conducting tournaments, lack of piety, etc.) as causes of the plague, and various attempts to appease God’s wrath, were recorded by chroniclers. It is estimated to have killed 30–60 percent of the population of Europe, the higher percentages being in cities. The epidemic finally burned out in 1351, only to be followed by another major plague epidemic in 1361, and yet another in 1374. That recurrence led to the sequestration of individuals: quarantine. Frith (2012: 5–6) describes that defensive stance:

> In 1377 the republic of Ragusa on the Adriatic Sea (now Dubrovnik in Yugoslavia) established a ships’ landing station far from the city and harbour in which travellers suspected to have the plague had to spend thirty days, the *trentena*, to see whether they became ill and died or whether they remained healthy and could leave. The *trentena* was found to be too short and in 1403 in Venice, travellers from the Levant in the eastern Mediterranean were isolated in a hospital for forty days, the *quarantena* or *quaranta giorni*, from which we derive the term quarantine. ... In the 14th and 15th centuries following, most countries in Europe had established quarantine.

Ancient DNA from archaeological sites supports the global distribution of the Black Death and subsequent outbreaks of plague. Over the fourteenth through sixteenth centuries, the plague devastated eastern and western Europe, Russia, Asia, the Mediterranean basin and the Middle East, and perhaps Southern Africa (Green 2014, 2018; Spyrou et al. 2019; and others). Outbreaks continued through the eighteenth century. The demographic effect varied from place to place depending on how often the plague returned. After the Black Death pandemic, epidemics returned to most of western Europe about every ten years and this had the greatest demographic impact, preventing populations from recovering their losses (Frith 2012).

Yersinia pestis has an unusually broad range of reservoir hosts (animals that can maintain an infection and serve as a source of the disease organism) in its transmission cycle, and the black rat–oriental rat flea model of plague transmission is no longer sufficient to explain the spread of the Black Death and recurrent outbreaks. There are substantially more vector-host pairings than were known in the late nineteenth century when Alexander Yersin discovered the bacterium and Paul-Louis Simond found it in rats and rat fleas.[4] Plague is primarily a disease of small, burrowing mammals, and especially rodents. Some of its most important reservoir hosts include the black rat (*Rattus rattus*), Norway rat (*Rattus norvegicus*), great gerbil (*Rhombomys opimus*), multiple species of marmots (*Marmota* spp.), Asia's plateau pika (*Ochotona curzoniae*), prairie dogs (*Cynomys* species), and their predators (Eisen and Gage 2012; Eads and Biggins 2015). Camels, sheep, and goats can also be reservoirs of plague in areas where rodents and their fleas are also infected (Frith 2012), and dogs and cats can carry it into the home. Over forty species have been identified as known plague reservoirs.

Over thirty species of fleas have been experimentally confirmed to transmit *Yersinia pestis,* with reports of up to 250 flea species producing naturally acquired infections (Eisen and Gage 2012: 66–9). When the disease begins an outbreak it moves from enzootic (endemic in animals) to epizootoic (epidemic in animals), and the number of host animals infected increases. Likewise when it moves from an epizootic to the zoonotic phase (transmitting to humans), numbers of infected individuals and species increase and are often in domestic spaces. While the black rat (*Rattus rattus*) was the likely reservoir when the Black Death arrived in Italy and its fleas (*Xenopsylla cheopis*) would have been the most likely vectors, the brown rat (*Rattus norvegicus*) and possibly other rodents could easily become reservoirs. Given the number of potentially suitable reservoir hosts and potential flea vectors, as well as unique conditions of the landscape, vegetation, geology, and climate across Europe and elsewhere, it is necessary to look for other links between reservoirs and insect vectors. Rates of spread of the Black Death and conditions such as the scarcity of rats in northern Europe and Scandinavia (Hufthammer and Walløe 2013) suggest different patterns of transmission as well as the involvement of human fleas (*Pulex irritans*)[5] and body lice (*Pediculus humanus humanus*), which can also transmit plague (Drancourt et al. 2006 and others). Dean et al. (2018) compared models of data from the plague from the fourteenth–nineteenth centuries using a variety of factors for human–human ectoparasites, human–human pneumonic transmission, and for rat–flea transmission. They found the best fit and more explanatory power for the human–human ectoparasite model applied to plague, including the Black Death, during this period.

Plague in humans takes one (or more) of three major forms, as well as a much less common form: 1) bubonic plague, 2) septicemic plague, 3) pneumonic plague, and 4) a gastrointestinal form, relatively rare and acquired by ingesting the bacteria. Typically the majority of cases are bubonic, the bacteria invading and replicating in the lymph nodes, which become painfully swollen "buboes" that can burst. Bubonic plague can also progress into pneumonic plague. According to the World Health Organization (2017), the bubonic form has a roughly 50 percent mortality if untreated, while the pneumonic form, which can be airborne in droplets and so is easily transmissible between people, and the septicemic form (circulating in the bloodstream) both have a 100 percent mortality rate if untreated. Antibiotics, effective today in treating many plague cases, were of course unknown during the Renaissance. The massive mortality in Renaissance plague outbreaks, all in "untreated" victims, was inevitable.

INSECT-VECTORED DISEASES: LICE AND TYPHUS, TRENCH FEVER, AND RELAPSING FEVER

Abundant evidence tells us that louse infestations were common during the Renaissance. There is less clear evidence for the occurrence of louse-borne diseases during that time, although it is now well known that lice transmit three major diseases: epidemic typhus, trench fever, and relapsing fever. These three well-documented louse-borne diseases are difficult to differentiate based on symptoms alone. Epidemic typhus is caused by *Rickettsia prowazekii*. The most common symptoms are fever, headache, and body aches. Cholewinski et al. (2017: 87) describe it thus: "*Rickettsia* bacteria trigger pathological changes of blood vessels, the nervous system, endocrine glands and the heart. Symptoms include fever, a feeling of helplessness, quickened breathing, migraine, swellings, enlarged liver and spleen, possible impaired consciousness and insomnia." Additionally, typhus produces a rash in less than half of the cases. Patients with relapsing fever, caused by the spirochete *Borrellia recurrentis*, can have an irregular, recurring fever, hence the name. Trench fever, caused by *Bartonella quintana*, does not have distinctive symptoms, which explains why it was not characterized until the First World War, when outbreaks occurred in the trenches of battlefields. Importantly, the disease-causing bacteria are found in the feces of the louse for all three infections, and the bacteria enter the body through crushing the louse into an abrasion, louse fecal contamination of scratches or breaks in the skin, or aerosolized dust. Considering that malaria was also present in most of Europe and the Mediterranean region and manifests as fever in some form, written descriptions of fevers become very difficult to sort out, and so the louse-borne fevers cannot be diagnosed from those descriptions with any precision during the Renaissance. However, aDNA evidence of both trench fever and typhus have been found, clarifying the occurrence of those diseases.

Epidemic typhus is the most serious infectious disease to depend on lice for transmission. Typhus is considered a disease of crowds and filth, which explains its other common names: jail (gaol) fever, camp fever, ship fever, and famine fever. Modern typhus outbreaks still have a mortality rate of 10 to 15 percent, though pre-antibiotic rates are believed to be much higher, up to 60 percent (Raoult et al. 2004). In a few cases, survivors can become long-term carriers of *Rickettsia prowazekii,* becoming reservoirs for the instigation of future epidemics (Burns et al. 2014). The only other animals known to be infected with *R. prowazekii* are the North American flying squirrel and its parasites.

Epidemic typhus was known in Europe during the Renaissance. “The first account of typhus was recorded during the civil wars of Grenada 1489–90, when typhus was introduced in Spain by soldiers from the island of Cyprus” (Socolovschi and Raoult 2009: 103). The first major epidemic occurred in 1527–1528, first from Naples to Milan, Italy and then spread northward through Europe as a series of small outbreaks. It was known as spotted fever in the sixteenth century in England, and the outbreak of 1557–1559 reportedly killed 10 percent of the population in that country (Socolovschi and Raoult 2009). The disease was described by the Italian polymath Girolamo Cardano in his book *De Malo recentiorum medicorum Ursu Libellus*, and by the Italian physician Girolamo Fracastorius in 1546 in a treatise *De Contagione et Contagiosis Morbis* (Bechah et al. 2008: 417; Socolovschi and Raoult 2009). Fracastorius distinguished typhus from plague (Bechah et al. 2008: 417), and described the typhus epidemics that occurred in Italy in 1505 and 1528. His description of an outbreak fits epidemic typhus as clearly as possible:

> fever and fatigue followed by general prostration and delirium in the first few days; then the body erupted into a general rash that looked like flea-bites or, in some cases spots about the size of lentils, the characteristic symptom that gave it its contemporary Italian name, *mal di petecchie* or ‘spotted sickness’ ... [this] telltale sign, a rash of red spots about the size of lentils, has led many historians to confuse it with plague.
>
> (qtd. in Eamon 2007: 159)

The series of outbreaks between 1489 and 1552 killed an estimated 78,000 soldiers (Socolovschi and Raoult 2009: 104).

There is some question of the origin of typhus, given that all of the early European epidemics seem to trace back to Spain and Spanish military activity in Europe, including the 1528 epidemic in Italy. The Spanish were also making regular trips to New Spain in the Americas during the early sixteenth century, and it is possible that *Rickettsia prowazekii* originated there.

> ... historical evidence suggests that typhus fever existed among South American natives in pre-Columbian days and recognizable epidemics

> occurred in Mexico before the arrival of Cortez at Vera Cruz in 1519. Moreover, there is a legend, credited by Bernal Diaz and by Nicolas Léon, that typhus destroyed the Toltec city of Tollan in 1116 AD. The description of typhus fever in Mexico was done by missionaries in 1545. Toward the end of the sixteenth century, typhus killed more than 2 million native Indians in the Mexico highlands. It is tempting to speculate that typhus was born in the chance meeting of an American *rickettsia* and a Spanish louse.
>
> (Socolovschi and Raoult 2009: 103–4)

It is also interesting to note that Pedro Menéndez de Avilés (1519–1574), founder of the mission of Saint Augustine, Florida in 1565, died from the earliest described case of typhus in Santander, Spain (Beding 2016: 464). However, other authors state that typhus was not recorded in the New World until 1655, and the first epidemic was recorded in Mexico where it was known as *tabardillo* (Burns et al. 2014: 442).

In many cases, typhus could not be distinguished from louse-borne relapsing fever (LBRF) until modern times. The LBRF bacterium, *Borellia recurrentis*, a spirochete, is also transmitted by the human body louse, *Pediculus humanus*. It produces a vague set of symptoms: fatigue, pain, a cough, anorexia, and a high fever. Yellowing of the skin (jaundice) occurs as the liver and kidneys are involved, and the spleen may rupture. Victims may bleed through the skin or mucus membranes. Progressively weaker relapses occur every seven to ten days (Badiaga and Brouqui 2012: 335). Today, the mortality rate for untreated louse-borne relapsing fever ranges from 10–40 percent (Badiaga and Brouqui 2012: 335; Guellil et al. 2018). Mortality rates peaked in the winter when people wore more clothes and clustered indoors (Hays 2005: 240).

Louse-borne relapsing fever is suspected as a pathogen in medieval and Renaissance famines, but it is difficult to definitively recognize from written records. In 2018, the LBRF genome was isolated from ancient DNA in the fifteenth-century remains of a woman in a double burial with a child at St. Clement's/St. Nicolay church cemetery in Oslo (Guellil et al. 2018). This disease might also have been a co-infection that contributed to mortality from other diseases during the Renaissance.

Trench fever, the third louse-borne disease, is caused by the bacterium *Bartonella quintana*. Trench fever is a debilitating disease that could have contributed to overall mortality as a co-morbidity, a secondary cause of death. The mortality rate is so low that there are no statistics upon which to base an estimate for the past. Symptoms of trench fever include shin pain, fever, dizziness, and headache which, like LBRF, recur every few days (Badiaga and Brouqui 2012: 335). The discovery of ancient DNA of *Bartonella quintana* came as a surprise to paleogeneticists. After finding trench fever aDNA in mass graves of soldiers from more recent periods (Raoult et al. 2006; Nguyen-Hieu et al. 2010), researchers examined remains in a mass grave dated between the eleventh and the fifteenth century in the village of Bondy, France. They found DNA of both *Yersinia pestis*, the plague, and *Bartonella quintana*, trench fever (Nguyen-Heiu et al. 2011). To what extent, if any, these diseases interacted is unknown.

INSECT-VECTORED DISEASES: MALARIA, BAD AIR, AND MIASMAS

Malaria is a mosquito-borne disease known from ancient times, was described by Hippocrates, and was recorded by historians and physicians of the Roman Empire.

> The story of malaria in Imperial Italy is drawn from a rich historical narrative ... Hippocrates' *On Epidemics*, Celsus' *De Medicina*, or Galen's *De Morborum Temporibus*. These texts describe the classic fever periodicity — tertian, semi-tertian, quartan or quotidian — that have been documented since the 5th century BCE.
>
> (Marciniak et al. 2016: 1)

The pattern of symptoms, regularly recurring paroxysms, makes malaria more recognizable, and it was common enough in ancient societies to gain the attention of physicians and historians. The name itself comes from the Italian *mala aria* meaning "bad air," as the cause was thought to be foul air around marshes and swamps; this was consistent with the theory that diseases were caused by miasmas. In the fifteenth century, Leonardo Bruni used "mala aria" in his *Historia Florentina:*

> After the Florentines had conquered this stronghold, after putting good guardians on it they were discussing among themselves how to proceed. For some of them it appeared most useful and necessary to reduce the army, more so as it was extremely stressed by disease and *mala aria* (bad air), and due to the long-lasting and difficult camps in unhealthy places during the autumn.
>
> (trans. qtd. in Hemplemann and Krafts 2013: 3)

Malaria occurred across Europe as well, and was endemic in many areas, including marshlands in England and in Italy, France, Germany, Switzerland, and elsewhere (Dobson 1989; Reiter 2000; Newfield 2017). In those localities it was important demographically (Franklin 1983; Newfield 2017). Dobson (1989: 3) noted of the English marshlands (e.g., Essex) in the sixteenth century that "Residents and visitors described the 'marsh fevers' of these districts, variously calling them 'agues', 'tertian fevers', 'quartan fevers' or 'intermittent fevers' and attributing them to the noxious vapours of the stagnant marshes." Detailed descriptions of "ague" were made by physicians such as William Harvey (1578–1657), who recorded his observations of patients and post-mortem observations of those who succumbed to the disease. He himself had suffered from "a persistent ague (that) had caused him to miss much of his final year (1598–99) at the University of Cambridge" (Reiter 2000: 5). Tertian and quartan fevers were familiar enough to make their appearance in literature as well: Dante (1265–1321) refers to a quartan fever, one of the forms of malaria, in *The Inferno*. Shakespeare mentioned agues in eight of his plays, incorporating symptoms of malaria (chills and shaking) into the stories portrayed in those plays.

The importance of malaria during the Renaissance may also be attested to by the large number of antimalarial remedies described in herbals of the time; for example, over three hundred plant taxa were described and/or included in recipes in six German herbals from the sixteenth century (Adams et al. 2011). Many of these recipes include quinine-containing plants or, in the late sixteenth century, Peruvian bark (Hempelmann and Krafts 2013); quinine is an antimalarial drug that is still used in various forms today.

Malaria in humans is caused by several species of parasitic apicomplexan protozoans in the genus *Plasmodium*, the three most important being *Plasmodium falciparum*, *P. vivax* (both tertian forms), and *P. malariae* (quartan malaria).[6] Development of an infection produces the sequence of: 1) chills and violent shaking, 2) a hot stage (fever), and 3) a sweating stage as the body temperature drops; this sequence is termed a paroxysm. Individuals with tertian forms exhibit a paroxysm every third day, while quartan malaria produces a paroxysm every fourth day.

Plasmodium has a complex life cycle requiring two hosts, an intermediate host in which asexual reproduction of the parasite occurs and a definitive host in which sexual reproduction of the parasite occurs. Female mosquitoes in the genus *Anopheles* are definitive hosts for malaria, while a mammal or bird serves as an intermediate host; of over four hundred species of *Anopheles* mosquitoes that have been described, about seventy species are capable of transmitting malaria to humans (Sinka et al. 2012). The parasite also exists in two main forms, the sporozoite and the merozoite, each of which has mechanisms to evade the host's immune system. Various strains of some malaria species are also tightly linked to their mosquito vector's feeding behavior, appearing at a higher concentration in the peripheral blood of the host during the period when the mosquito is most likely to feed. This adaptation makes it more likely that malaria parasites will be picked up by a mosquito, whether the species/strain is a day-biting or night-biting mosquito (Roberts et al. 2013).

To summarize the *Plasmodium* life cycle, infected female mosquitoes inject the malaria parasites (sporozoites) into the blood of a human (or other) host when they take a blood meal. The sporozoites travel in the bloodstream to the liver, where they invade liver cells, consuming the cell contents and undergoing cell division to produce more of themselves but in a new form, the merozoite. The merozoites then burst out of the liver cells (although in *P. vivax*, some parasites can remain in the liver in a dormant state) and enter the bloodstream, attacking red blood cells (erythrocytes) in the disease-causing phase known as the erythrocytic cycle.

A single merozoite entering a red blood cell will consume the cell contents, producing a number of waste products and toxins, and it will divide to produce eight or more replicates of itself. These break out of the red blood cell and each new merozoite attacks another red blood cell, beginning a new erythrocytic cycle. Paroxysms are the host's response to the simultaneous bursting of red blood cells and release of toxins that occurs at the end of each cycle, three days in tertian malarias and four days in quartan malaria. In *Plasmodium falciparum*, the bursting of red blood cells is not as synchronous as it is in other malaria species, so that paroxysms may be almost continuous through the three-day cycle, the fever dropping only for a short time every third day (Roberts et al. 2013). There are other consequences that may occur in a malaria infection as well, including liver and spleen damage, brain damage, and kidney damage. Hemozoin, a black waste produced from the breakdown of hemoglobin, can accumulate in tissues and causes the form of malaria known as "blackwater fever."

Malaria is unique among insect-vectored diseases; unlike other diseases in which the pathogen multiplies (bacteria and viruses) and perhaps even changes form (trypanosomes) during the transmission cycle, malaria has a life cycle including sexual reproduction, and that life cycle cannot be completed without the mosquito. As parasites are multiplying in the bloodstream of mammalian or avian hosts, some produce gametocytes, cells that eventually will produce gametes. These remain circulating in the blood, and when a mosquito takes a blood meal, gametocytes are also taken up; approximately a thousand gametocytes are picked up in an average blood meal (Sinden 1999). Once in the mosquito, these cells produce gametes and fertilization occurs, producing embryos in the sexual phase of the life cycle. The embryos will eventually produce many new infective sporozoites, which travel to the mosquito's salivary glands and penetrate them, ready to be injected into the next host that the mosquito bites.

While malaria seems to have been a constant in Europe since ancient times, it apparently did not exist in the Americas until the exploration, conquest, and colonization of the early sixteenth century. The *Plasmodium* parasite was first introduced to Hispaniola, assumed to be the site where Columbus first landed,[7] by the Spanish explorers (Cook 2002: 375–6). Malaria had to come from human reservoirs, probably persons on board ships from Europe. In the late sixteenth century, ships filled with slaves bringing the parasite with them from its African origin may have made falciparum malaria infections more common. Yet, if the African slave trade brought the bulk of the introductions, it was the Europeans who first brought malaria to the Caribbean and the Americas. Malaria invaded the Americas relatively easily because *Anopheles* mosquitoes capable of acting as hosts/vectors were already well established (McNeill 2010).

European conquest of the New World added another dimension to the impact of insects on human health. Before contact with Europeans, there were no known mosquito-transmitted diseases in the New World. Two significant events that would impact disease in the New World's inhabitants occurred before 1600: the introduction of malaria to the colonies established by European nations, and the introduction of the mosquito *Aedes aegypti* from Africa to South and Central America. The yellow fever virus and its vector, *Aedes aegypti*, originated in Africa; the virus occurs non-human primates, as well as humans. That last introduction, of *Aedes aegypti*, set the stage for an Africa-to-Americas conveyor belt of viruses from yellow fever to zika that would in future centuries wreak havoc. Malaria and yellow fever would each play a role in shaping the Americas, but not until the seventeenth and eighteenth centuries. The late Renaissance set the stage, but the full impact would come in the next century (McNeill 2010).

INSECT-VECTORED DISEASES: VISCERAL AND MUCOCUTANEOUS LEISHMANIASIS

Visceral leishmaniasis is a disease caused by two species of kinetoplastid protozoans, *Leishmania donovani* and *Leishmania infantum*; visceral and a cutaneous form occur in the Old World and in Italy are caused by *L. infantum* (Nerlich et al. 2012). Mucocutaneous leishmaniasis, caused by *L. tropica*, *L. braziliense*, and others, is an ancient disease of the Americas. These kinetoplastid protozoans are transmitted from infected hosts, including humans and dogs, by blood-feeding sandflies in the genera *Phlebotomus* (Old World) and *Lutzomyia* (New World). Leishmaniasis caused by *L. infantum* has occurred in the Mediterranean region, as well as Africa, since antiquity, having been found in mummies from Egypt and Nubia (Zink et al. 2006), while *L. donovani* has long plagued people in Asia (Roberts et al. 2013).

The peoples of South and Central America also were, and are, victims of the mucocutaneous form that is very disfiguring and can destroy the nasopharynx and tissues. The existence of that disease is depicted in pre-Columbian ceramic figurines of disfigured people, along with figurines of healthy individuals, from the Peruvian Andean people (Novo et al. 2016), evidence of its existence and its impact on culture in the Americas since before the European conquest of the New World. Additional written descriptions are found in two Spanish reports:

> … early reports of the disease in South America have been published by the Spanish Balthasar Ramirez in his book "Descripción del Reyno del Peru" (1580), where the disease was called "Mal de los Andes". In the same period, Fray Rodrigo de Loayza (1586) noted the presence of nasal lesions not only in Indians but also in people entering the Andes region.
>
> (Novo et al. 2016: 45)

The presence of *Leishmania* aDNA in pre-Columbian mummies confirms that cutaneous and mucocutaneous forms of *Leishmania* were present in South America (Costa et al. 2009; Gerszten et al. 2012), although the species (singular or plural) in some is uncertain.[8] References to leishmaniasis also indicate its presence in the Renaissance period: "The notification of a disease named 'valley sickness' or 'Andean sickness' or 'white leprosy' is mentioned during the 15th and 16th centuries AD that correspond to the Inca period" (Sereno et al. 2017: 3). However, *Leishmania infantum*, the species causing visceral leishmaniasis, was brought to Central America in the sixteenth century by Europeans and their infected dogs. All of the forms of the disease persist in Central and South America to the present day.

The cause of visceral leishmaniasis was unknown through the Renaissance, just as was true for bacterial diseases and other protozoan parasitic diseases; miasmas were thought to cause this and virtually all other illness. However, illness and other health conditions were observed and recorded for individuals in positions of power. That was the case for Eleonora of Toledo, wife of Cosimo I de' Medici, Grand Duke of Tuscany, in the sixteenth century. Eleonora had declining health through her thirties and died at the age of forty; reports of court physicians as well as correspondence and other archival sources indicate that "during the final twelve years of her life, Eleonora suffered from several severe ailments. Irregular bouts of fever, wasting and constant vomiting, stomach pain, weight loss, anemia, persistent coughing, and hemorrhage from the mouth …" (Bianucci et al. 2012: 233). While these symptoms suggest tuberculosis, which was confirmed at autopsy by lesions in her lungs, she and other family members had gone to marshy areas of Tuscany on a hunting trip shortly before her death, where she could have contracted falciparum malaria. Two of her sons who had gone along to Tuscany also died at the same time as Eleonora.

Examination and genetic testing of her remains and also the remains of her husband and the two sons who died confirmed tuberculosis in Eleonora but not in the other family members. Surprisingly, falciparum malaria aDNA was not found, but aDNA from *Leishmania* was found, confirming that she had visceral leishmaniasis, which was known from Tuscany at that time. Eleonora's symptoms were consistent with untreated visceral leishmaniasis. This was the first instance of finding a co-infection by use of aDNA from archaeological remains (Bianucci et al. 2012; Nerlich et al. 2012). Further, immunological evidence of falciparum malaria had previously been found in Eleonora (Fornaciari et al. 2010), so the negative result for aDNA does not necessarily rule out malaria as a co-infection.

INSECT-VECTORED DISEASES: AFRICAN AND AMERICAN TRYPANOSOMIASIS

Trypanosomiasis has plagued peoples in Africa and in Central and South America for thousands of years; in fact, anthropologists and biologists have discussed trypanosomes in Africa as an important factor influencing the evolution of hominid species (Lambrecht 1985). Trypanosomes, species in the genus *Trypanosoma,* which are kinetoplastids, are responsible for very different diseases in the two hemispheres. Subspecies of *Trypanosoma brucei* in sub-Saharan Africa infect people and animals, causing sleeping sickness and nagana respectively; tsetse flies in the genus *Glossina* transmit the parasites. *Trypanosoma cruzi* causes Chagas disease in Central and South America, the trypanosomes being transmitted by blood-feeding bugs in the hemipteran family Reduviidae, the cone-nosed or kissing bugs. These diseases were generally unknown to Europeans in the fourteenth and fifteenth centuries, but were "discovered" following the conquest of the Americas, and with increased trade in Africa.

African trypanosomiasis takes two forms, each caused by different subspecies of *Trypanosoma* and transmitted by different species of *Glossina.* The chronic form today accounts for 98 percent of cases (WHO 2020); found primarily in West Africa, it initially is an infection of the blood and lymphatic system, the trypanosomes invading the blood and extracellular spaces and multiplying. One diagnostic indication of infection is Winterbottom's sign, a greatly swollen lymph node, often a cervical lymph node. The parasites eventually invade the central nervous system, which results in neurological symptoms including confusion, loss of motor coordination, sleepiness, coma, and, eventually, death. This form has a relatively slow progression and an infected person may live for years, with gradually degenerating function. The acute form is found in East and Central Africa. The trypanosomes invade the central nervous more rapidly, as well as affecting neuromuscular junctions in the heart and other organs, and death results in less than two years. Many different wild animals, as well as humans, act as reservoirs for both forms of the disease, and trypanosomiasis (nagana) in cattle and other animals can have devastating consequences. The "tsetse belt" effectively prevented cattle raising by Indigenous peoples to whom it was central to their culture and livelihood.

African trypanosomiasis during the fourteenth through sixteenth centuries was recorded by Arab traders who traveled within West Africa. One of the first descriptions of a specific case was written by the Arab historian Ibn Khaldun in the fourteenth century. He recorded a report of the illness and death of the Emperor of Mali, Sultan Mari Jata:

> He told me that Jata had been smitten by the sleeping illness, a disease which frequently afflicts the inhabitants of that climate, especially the chieftains who are habitually affected by sleep. Those afflicted are virtually never awake or alert. The sickness harms the patient and continues until he perishes.
>
> (qtd. in Steverding 2008)

The connection between tsetse flies and trypanosomes was not recognized until the end of the nineteenth century, as none of the earlier reports and descriptions of the disease included any reference to flies.

American trypanosomiasis, known today as Chagas disease, is an ancient zoonotic disease endemic in Central and South America. It is a sylvatic disease, meaning that it is found in forested habitats such as the Amazon basin. *Trypanosoma cruzi*, the etiological agent, is found in primates and hundreds of other forest-dwelling mammals, including sloths, armadillos, coatis, opossums, and bats (Jansen et al. 2017, 2018). The disease is transmitted by blood-feeding bugs primarily in the genera *Triatoma* and *Rhodnius*, as well as by ingestion of contaminated meat (Guhl 2017); this latter mode of transmission may in part explain the large number of reservoir species.

The trypanosomes are found in the digestive tract of a kissing bug, and are deposited on the bug's host in its semi-liquid feces. The parasites get into that mammalian host, including humans, when fecal material gets into a break in the skin such as the site of the bug bite, or gets rubbed into the eye.[9] Unlike their African congeners, once inside the host the trypanosomes invade cells of any tissue, change their form and multiply, and then burst out of the cells, each parasite invading another host cell. Once infected, the infection is permanent, and long-term cumulative damage to organs can be severe or fatal.

The disease in humans has two phases, an initial acute phase and a later chronic phase. The acute phase has symptoms typical of an infection, including fever and general discomfort, but it may be asymptomatic other than a lesion at the site of the point of entry or, if rubbed into an eye, swelling of the eyelid (the sign of Romaña). After a few weeks, the disease goes into a chronic phase, with the host becoming asymptomatic as the immune system adjusts to the parasite. The chronic phase is often asymptomatic, but some individuals (after ten or more years) develop the organ damage that is a hallmark of this disease, organs that increase in size but lose muscle function: cardiomyopathy, megaesophagus, and megacolon. Teixeira et al. (2006) report these changes to be associated with immune system lysis of parasite-free cells by inflammatory cells such as lymphocytes. A low level of parasitemia remains in the blood, enabling diagnosis using sterile kissing bugs to collect and incubate the trypanosomes if they are present (xenodiagnosis).

This disease, like malaria and others, is an ancient one, and so was certainly present in its endemic areas through the Renaissance. *Trypanosoma cruzi* aDNA has been found in human mummies and remains in northern Chile and southern Peru. Based on aDNA, Steverding (2014) places the prevalence of *T. cruzi* infection at 41 percent in this area at the time of the Spanish conquest in the sixteenth century. Additional evidence of pre-Columbian *T. cruzi* infection comes from thirteenth–sixteenth century Peruvian ceramic figures that depict manifestations of Chagas disease, including single swollen eyelids (Stevarding 2014). He also notes that transmission of Chagas disease is facilitated by deforestation and more open landscapes; the vectors prefer such habitats, and have become adapted to living in human habitations as well. This is a disease of very complex interactions among reservoirs, vectors, habitats, and ecosystems.

A LAST WORD ...

The frequency of co-infections in various populations during the Renaissance can never be fully known. That they existed has been demonstrated by archaeological investigation of human remains at a variety of sites in Europe, the Americas, and elsewhere. That they were common seems highly likely. Plague and anthrax, plague and malaria, malaria and trench fever, typhus and plague, leishmaniasis and tuberculosis and malaria—these are just a few of the possible combinations. Interactions between disease organisms, especially the suppression of immune responses, may have enhanced the pathologies of both (or all) infectious organisms in a co-infection. Continuing analyses of DNA and immunological proteins in known and yet-to-be-discovered archaeological sites and from human remains will no doubt shed more light on the diseases that afflicted people during the fourteenth through sixteenth centuries.

Insects played a major role in the transmission of diseases, not to mention those insects that infested people and at best caused irritation and discomfort through bites. Other impacts of insects on individuals we will never know: allergies to bee stings, or insect phobias for example. However, the consequences of infestation and disease transmission by insects spanned all aspects of human culture: health and mortality, demographic shifts, economic gains and losses, artistic representations of death, social structure, changes in faith and belief systems, and literature. Insects played crucial roles in the disease landscapes of the Renaissance, as they continue to today.

CHAPTER THREE

Insects and Food

MICHAEL BLUST, E.M. ALDOSORO MAYA, AND
EVANGELINA BLUST

The period 1300 to 1600 is generally referred to as the Renaissance, although the term "late Middle Ages" is considered to include up to around 1500. The scientific revolution was beginning late during this time frame and the printing press first appeared halfway through this period. Particularly pertinent to the insect world, the "Age of Exploration" includes the latter half of this period. It is important to keep in mind that taxonomy was still in its very early stages and humans were not entirely consistent as to what was termed an "insect," especially when considering immatures and other worm-like creatures. Scientific analysis of cause and effect was rare, as was the understanding of ecological relationships. While the big insect-related story of this period was the Black Death, insects played a variety of roles affecting the diet of humans, especially Europeans.

AGRICULTURAL PESTS

Agricultural systems during this time largely consisted of species native to an area. Travel had not progressed to the point where introduction of pests and consequent problems with invasive species were significant. While many crops had typical problems with pest species, large-scale insect devastation was rare. One notable exception to this pattern were locust plagues. They were recorded in biblical times and before, so they were not new to the Renaissance, but their frequency and intensity made them the most recorded agricultural pest during this era. Camuffo and Enzi (1991) analyzed a wide variety of sources and compiled documentation of locust invasions affecting Italy in particular, and Europe in general, from the Middle Ages to 1800. The fourteenth and sixteenth centuries were the most affected; about seventy of these invasions were documented, three of which are given here:

> 1365 May/June (Libro di commentarii, Cod. Clasense 108, vol. II)
> During the same year 1365, in the beginning of May there was (in Ascoli area) a great number of crickets, damaging all the wheat, legumes and vegetables, causing shortage of everything.
>
> 1478 Summer (Cronica gestorum, p. 27)
> At this time an exceptional event occurred in the countryside around Mantova. Grasshoppers in fact appeared which for many days ate grass, fruit, vines and all the vegetables, and they were so numerous that they covered the land from the Brescia side for thirty miles (53.5 km) in length and for four miles (7.15 km) in width. Moreover, they deposited their eggs, from which, if rubbed between the hands with earth or dust, immediately small locusts called "grasshoppers" were born.
>
> 1573 (Langio, pp. 446–53)
> And this terrible misfortune and wonder happened in Egypt in 805 before the foundation of the city (Rome). However, it is better to remember the more recent cases. Since that year, when we mentioned these things, again in (15)73 grasshoppers flew in swarms into Italy, destroying around Brescia and Mantova all the shoots in the soil as far as the roots, with their insatiable hunger: following this there was a plague, which killed more than fifty thousand people. And you, Lisandro, were born a long time before me, you are not unaware that twenty-four years ago an immense swarm of grasshoppers flew from Tracia to Poland, Hungary and as far as the borders of Germany; which not only devoured the forage but also vegetables and every other green thing in the fields and meadows, beyond measure, which led some to forcast extraordinary and incredible events from the coming of the cruel army of the Turks. This presentiment was justified, in fact immediately the following year there was the siege of Vienna, the devastation of Hungary, and from the land of the Angles and the Britons that mortal fever accompanied by perspiration reached us: in fact that pestilential epidemic not only of the trees but also of man was originated by the corruption of the air and the bad weather ...
>
> (Camuffo and Enzi 1991: 59, 60, 64)

Given the wide variety of sources compiled, the terminology for units of measurement, insect species, and life stages, are inconsistent, but give valuable information nonetheless.

In similar meta-analyses of locust infestations in Asia, Tian et al. (2011) and Stige et al. (2007) examined historical records from local gazetteers going back over thousands of years. In China, locust plagues had long been documented by most local governments in hopes of finding predictive patterns. While these analyses do not enumerate the outbreaks, graphical presentations show average to above average frequency of outbreaks during the period from 1300–1600, and the sixteenth and seventeenth centuries are mentioned as a period of peak locust abundance. Camuffo and Enzi (1991) also noted the sixteenth century as one of the worst locust periods in Italy. The data from the sixteenth century on showed that “there were more locusts under dry and cold conditions and when locust abundance was high in the preceding year or decade” (Tian et al. 2011: 14521). Tian et al. (2011) further note that cooling conditions tend to reduce the monsoonal rainfalls that may negatively affect locust populations. However, the situation is complex in that conditions in one region may be good for breeding, followed by good migratory weather conditions, and then good survival weather in the region subject to the invading plague. Stige et al. (2007) note that receding floods and receding rivers due to drought both leave good soil conditions for oviposition and egg survival. Camuffo and Enzi (1991) found that cold windy conditions often correlate with the end of a locust outbreak.

In the mid-sixteenth century, Leo Africanus described locust plagues in Africa that were so huge, they cast a yellow glow in the sky a day in advance of their arrival. He states:

> Of locustes there are sometimes seen such monstrous swarmes in Africa, that in flying they intercept the sunne-beames like a thick cloude. They deuoure trees, leaues, fruites, and all greene things growing out of the earth. At their departure they leaue egges behinde them, whereof other yoong locusts breede, which in the places where they are left, will eate and consume the things euen to the very bark of trees, procuring thereby extreme dearth of corne, especially in Mauritania. Howbeit the inhabitatnts of Arabia deserta, and of Libya, esteeme the coming of these locusts as a fortunate boading: for seething or drying them in the sun, they bruise them to powder, and so eate them.
>
> ([1600] 1896: 957–8)

In the following passage, Africanus quotes from the writings of Francisco Alvarez, a Portuguese missionary in Ethiopia from 1515–1521:

> In this quarter and throughout all the dominion of *Prete Ianni*, there is an horrible and great plague, to wit an innumerable companie of Locustes, which eate and consume the corne, the trees of fruite; and so great is the number of these creatures, as it is not credible, for with the multitude of them the earth is couered, and the aire so ouerspred, as one my hardlie

> discerne the sunne … and wheresoeuer they come, the earth is more wasted and destroied by them, then if it had beene all ouer consumed with a fire. These locusts are as bigge as the greatest grashoppers, hauing yellow wings. Their coming into the countrie is knowne a day before: not for that we can see them, but we know it by the sunne, who is yellow of colour, this being a signe that they draw neere to the countrie, as also the earth looketh yellowe, by reason of the light which reflecteth from their wings: whereupon the people in a manner become presentlie halfe dead, saying, we are vndone, for the Ambati, that is to say, the locusts are come.
>
> (Africanus [1600] 1896: 961–2)

Alvarez's appeal to God for deliverance from the locust plague was successful, although his excommunication of the locusts was not:

> When this scourge and plague was come, the priestes of that place came and sought me out, requesting me to giue them some remedy for the driuing of them away, and I answered, them that I could tel them nothing, but only that they shoulde deuoutly pray unto God, that he woulde driue them out of the countrie. And so I went to the Ambassadour, and told him, that it would be very good to goe on procession, beseeching God that hee woulde deliuer the countrie, who peraduenture in his great mercie might hear us. This liked the Ambassadour very well: and the day following we gathered together the people of the land, with all the priests, and taking the consecrated stone, and the crosse, according to their custome, all we Portugals sung the Letanie, and appointed those of the land, that they should lift up their voices aloud as we did … and with this manner of inuocation we went ouer a peece of grounde, where there were fieldes of wheate, for the space of a mile, euen to a little hill: and here I caused many of these locusts to be taken, pronouncing ouer them a certaine coniuration, which I had about me in writing, hauing made it that night, requesting, admonishing, and excommunicating them, enjoining them within the space of three howers to depart towards the sea, or the lande of the Moores, or the desert mountains, and to let the Christians alone: and they not performing this, I summoned and charged the birdes of heauen, the beasts of the earth, and all sorts of tempests, to scatter, destroy, and eate vp their bodies … In the mean while towards the sea, there arose a great cloude with thunder, which met them full in the teeth, and continued for the space of three howers with much raine, and tempest, that filled all the riuers, and when the raine ceased it was a fearefull thing to behold the dead Locustes, which were more than two yardes in height vpon the bankes of the riuers, and in some riuers there were mightie heapes of them, so that the morning following there was not one of them found alive vpon the earth.
>
> (Africanus [1600] 1896: 962–3)

Other than such dramatic accounts of locust plagues we have few records regarding insect pests on crops. Curran (1937) stated that the average Aztec of the sixteenth century was well acquainted with insects that would later be major crop pests for the European settlers in the Americas, including the corn earworm. This does not mean, however, that the Aztecs viewed these insects as pests. They ate both the corn and the corn earworms (*Helicoverpa zea*). The earworms were a source of protein that had been raised on corn—not much different from beef cattle today.

PEST CONTROL

Crops have always had insect pests, and it seems likely that humans have always tried to reduce losses of crops to insects. Early information on agricultural pests was more likely to be recorded when the pests reached levels that were damaging; that was when growers focused on control of the pests. Both chemical and biological controls of insects did not really progress until after the scientific revolution itself had advanced.

Thacker (2002) noted that agricultural methods in Europe stayed relatively unchanged for hundreds of years. Use of botanicals as insecticides significantly expanded when explorers to the New World observed native Americans in Venezuela using powdered seeds to protect crops. The seeds came from a lily, *Sabadilla officianarum*. This plant was imported to Europe and raised for the extract that was used, and is still used in some places, for control of insect pests. In addition to *Sabadilla*, nicotine's effectiveness as an insecticide accompanied the discovery of tobacco by the explorers, and subsequently nicotine was used in Europe and the colonies. Koul and Dhaliwal (2001) reported the use of false hellebore (*Veratrum album*) in Europe to make hellebore powder which was a powerful insecticide. *Sabadilla* and hellebore both possess alkaloids which affect sodium channels of the nerves. These alkaloids are toxic to insects, and equally toxic to mammals. Emil Mrak (1969) noted that Marco Polo purportedly brought pyrethins from China to Europe, but it was unclear if they were used in agriculture, as their primary recorded purpose was for controlling lice.

Starna et al. (1984) described the insect pests affecting crops of the Iroquois in North America, especially corn, beans, and squash. Starna's work included a degree of speculation, but much of the Iroquois lore handed down through time described creatures attacking crops, indicating that it was a significant issue for the Iroquois. Native tales and notes from early explorers suggest that cutworms, grasshoppers, wireworms, cicadas, and possibly corn rootworm were some of the probable pests. The Iroquois probably had few options for control of crop pests. Spring tillage might have been somewhat useful for cutworm control, but tilling was not easy for the partially cleared woodlands that were typical fields for the Iroquois. Crop rotation was practiced, but it was questionable how much this was done as a form of pest control versus as a response to depletion of soil fertility. "Corn medicines" might have been used to protect seeds and seedlings. These were aqueous extracts of certain plants in which corn seeds were soaked before planting. One of these plants, a North American species of hellebore (*Veratrum viride*), has been shown to have insecticidal properties. It also seems likely that insect pests might have played a role in the periodic abandonment of villages and relocation to other sites (Starna et al. 1984).

Meanwhile, inorganic insecticides became more readily available and commonly used. The Chinese were reported to have used arsenicals as insecticides at least back into the sixteenth century and probably much earlier (Ecobichon 2001: 763). Arsenic, mercury, and lead were used as insecticides in the same period, mentioned in a report by the International Rice Research Institute in the Philippines. Needham in his multi-volume treatise *Science and Civilisation in China* made reference to the practice of sprinkling lime or tung oil on rice leaves as a deterrent to foliage feeders in the sixteenth century (1974: 506).

Herbs had many uses including insect control from Antiquity on through the Middle Ages. The development of the printing press helped spread knowledge of and maintain these uses during the Renaissance. While many of the uses were not directly related to protecting crops, the practice of "strewing" herbs on the household floor to discourage pests of all kinds was common. Tansy in particular was widely used to control vermin including mice, fleas, and flies. As such, it was also a common garden plant, but it is difficult to discern to what extent it was grown to protect vegetables in the garden. In the ninth century, Charlemagne had tansy in his garden (LeCain and Sheley 2014). If true, it was likely that during the Middle Ages people continued such practices. During the Renaissance, wealthy Europeans had access to spices which helped food to keep longer and repelled maggots, moths, and weevils (Albee 2014). The lower classes, on the other hand, had to tolerate a certain amount of spoilage in the food they ate.

Physical and non-poisonous forms of pest control were also used. Physical control includes crushing, barriers, and substances that interfere with breathing, moving, or reproduction of the pest. It is not unreasonable to assume that pests such as caterpillars would have been crushed by a person tending a garden or fruit trees during the Renaissance. In sixteenth-century Japan, a more elaborate practice was used: low quality whale oil was rendered so that it could be mixed with vinegar and sprayed onto rice crops as an insecticide. This mixture worked by interfering with the proper development of the insect cuticle (Ecobichon 2001: 763).

Biological control is the use of living agents (predators, parasites, or pathogens) to reduce damage by a pest species. One of the earliest records of biocontrol came from citrus orchards in China in the sixteenth century and perhaps before. Nests of the predatory ant species *Oncophylla smaradina* were placed within an orchard, and some reports stated that bamboo "highways" were constructed to connect trees for easier travel by the ants. These insects were used to control foliage feeding insects. The advent of chemical control replaced the ants and other biological control methods. Some efforts are now being made to re-introduce some of this older knowledge that has been lost by many farmers (Doutt 1964; Niu et al. 2014).

INSECT TRIALS

So how did one control insects when the number was beyond the scope of the meager methods at one's disposal? During the fourteenth and fifteenth centuries in particular, the Church and the justice system were frequently called upon to deal with situations where humans and other animals came into conflict. In 1891, Professor Karl von Amira used the terms *Thierstrafen* and *Thierprocesse* when referring to such dealings. *Thierstrafen* referred to the use of secular tribunals to issue capital punishments on domestic animals accused of homicide, whereas *Thierprocesse* referred to the use of ecclesiastical courts to mete out justice upon vermin that caused agricultural destruction. Unlike the larger animals, insects and other small species could not be physically controlled by humans. Metaphysical attempts at control were seen as the only viable option. Plagues were considered as either punishments from God or curses from the devil; either way, they fell into the domain of the Church. If the plague was a curse from God, then contrition and forgiveness were called for. If the plague was a curse, then it was time to ask God's protection and perhaps an exorcism of some sort. Another major difference between *Thierstrafen* and *Thierprocesse* had to do with size and number: domestic animals could be dealt with directly on an individual basis, but insects and other pests could not. A few representative individuals could be collected, but the objective was to try to make the whole group leave the area and/or stop the destruction (Evans 2009).

Bartholomew Chassenée of France was a sixteenth-century lawyer who built his reputation by representing pests in court. In 1531, he wrote the first of sixty-nine *consilia*, which were medieval versions of advice books, often dealing with medical issues and plagues. In this volume, he analyzed the case of a town's appeal to an ecclesiastical tribunal against "huberes" or "hurebers" (probably a locust or harvest fly). Using a series of courtroom and philosophical arguments, as well as previously written works, including the Bible, he concluded that the process was legitimate—animals can be tried and sentenced by both civil and

FIGURE 3.1 Depiction of the 1457 trial of a sow and her piglets for murder in Lavegny (sp), Switzerland, https://commons.wikimedia.org/wiki/File:Trial_of_a_sow_and_pigs_at_Lavegny.png.

ecclesiastical courts (Figure 3.1). Chassenée then cited a number of examples to demonstrate that anathemas (formal cursing or excommunication by the Church) were indeed effective. In addition to episodes involving plants, fish, and birds, he recounted a swarm of flies in the abbey church of Foigny that were excommunicated by St. Bernard. The flies were all dead by the next day. Such cursings from saints or directly from the Pope did not need legal proceedings in order to be enacted (Evans 2009).

It was argued by some medieval philosophers and theologians that animals had rights similar to those of humans, and therefore, efforts were made to enact proceedings that recognized these rights. However, the difficulties that accompanied trying an entire species, or at least an entire population of the species, resulted in rules and rituals that were rather contorted at times. For instance, when trying humans, rules applying to laity and clergy were different. If vermin were considered as equivalent to humans, then one of the first steps in the process was to determine if the vermin were laity or clergy. After some complex legal argumentation, Chassenée concluded that unless one could come up with a reason to consider them as clergy, the default position was that they were laity.

In some cases, the offending pests were given warnings and time to leave before the pronunciation of an anathema. For example, in 1516 in Troyes, an insect pest (probably a weevil) was attacking the grape vines. The pests were given six days to disappear before being cursed. One has to wonder whether this grace period was actually increasing the time for the threats and curses to appear effective by allowing time for natural control factors to come into play. Similarly, in 1545 a complaint was made to the Church regarding weevils attacking the vineyards of St. Julien in France. As an example of how convoluted these legal proceedings had become, it was recognized that God had made the fruits of the earth for all creatures, not just for humans, so the weevils had a right to feed. It was therefore decided that the residents must be guilty of something and should therefore take care to turn from any sinful ways and ensure that they were paying their tithes. After a couple of masses and a formal proclamation, the weevils disappeared and all was good for about thirty years. When the pests returned, a trial was initiated with a series of courtroom formalities, religious back-and-forth, and extenuating circumstances; it took eight months to bring the trial to a conclusion. It was unclear whether the weevils had the patience to wait. While we don't know exactly what the pest species was, most adult insects only live for a month or two.

In 1338, a locust plague developed near Botzen in Tyrol. Because the threat was growing rapidly, the ecclesiastical court heard the prosecution against the locusts and the parish priest was asked to excommunicate the beasts in accordance with the tribunal's verdict. The efficacy of anathemas was thought to be dependent in part on the exactness of ritual, but also on the faith and faithfulness of the affected population. Sinfulness in the local peoples and inadequate or delayed tithing reduced the effectiveness of the curse for a while, but the locusts ultimately succumbed to the power of the Church.

In 1541 a plague of locusts destroyed crops in Lombardy, Italy. Accounts of the disaster included the death of many people due to starvation. In such plagues, it was not uncommon that the grasshoppers would "eat themselves out of house and home" and many of them would starve as well. The vast number of dead insects was said to stink so strongly that vultures and other carrion eaters were repelled. The flowery language used in arguing these cases can be seen in an example from the opening statement of the prosecuting attorney:

> Gentlemen, these poor people on their knees and with tearful eyes, appeal to your sense of justice .... In the power of excommunication you have a weapon more effective than any wielded by that emperor [Augustus Caesar] to save these poor supplicants from impending famine produced by the ravages of little beasts, which spare neither the corn nor the vines .... It remains, therefore, after complying with the usual forms, only to adjudicate upon the case in accordance with the facts stated in the Petition of the

> Plaintiffs, which is right and reasonable, and, to this effect, to enjoin these animals from continuing their devastations, ordering them to quit the aforesaid fields and to withdraw to the place assigned them, pronouncing the necessary anathemas and execrations proscribed by our Holy Mother, the Church, for which your petitioners do ever pray.
>
> (qtd. in Evans 1906: 96)

There was also an account from fourteenth-century Germany of a community that brought forward a complaint against Spanish flies. The flies were cited to appear, but appointed a representative (curator) who defended them. Though not able to prevent their banishment, he was able to have some land given for the flies to retire to. In such cases, the pests were to leave the affected area and move. If they disappeared, the ban, and therefore the Church, was given credit. There is no information on whether the pests actually appeared on the land assigned to them. If the pests remained, it was the sins of the people that made the ban ineffective.

In another case, in 1478 in Berne, Switzerland, an insect referred to as an "inger" was brought to the attention of the Church. In this case, an interesting workaround was used in a case presented to the Bishop of Lausanne to avoid cursing God's creatures. Evans (1906: 114), by way of a Swiss writer named Schilling, quotes Bernhard Schmid, the local parish priest:

> Thou irrational and imperfect creature, the inger, called imperfect because there was none of thy species in Noah's ark at the time of the great bane and ruin of the deluge, thou art now come in numerous bands and hast done immense damage in the ground and above the ground to the perceptible diminution of food for men and animals.

This appeared to deny the inger its status as one of God's creatures. It was lower than the animals and was not even allowed on Noah's ark, thus clearing the way for it to be cursed without any guilt on the part of humans.

Cases of "Animal court" were documented from as early as 824 when some moles in northern Italy incurred the wrath of locals, to at least 1906 when a dog in Switzerland was condemned to death for its role in helping its owners in a murder-robbery. Table 3.1 summarizes insect-related trials from the ninth through the nineteenth century, as reported by Evans (1906).

The persistence of this practice of controlling pests via anathemas makes one wonder why such practices were viewed as worthwhile for such a long period of time. As Paul Robertshaw (1998) pointed out in his online discussion of medieval pest control, plagues of pests were unpredictable and their cause was not understood. We understand now that pest species are often subject to

TABLE 3.1 Chronological list of excommunications and prosecutions of insects from the ninth to the nineteenth century (adapted from Evans 1906).

Sources of Information	*Dates*	*Animals*	*Places*
Muratori: Rer. Ital. Scriptores, iii	886	Locusts	Roman Campagna
Sainte-Foix: Oeuvres, iv, p. 97, Mémoires de la Société Royale des Antiquaires de France, viii, p. 427	1120	Field-mice and Caterpillars	Laon
Théophile Raynaud: De Monitoriis in Opusc. missc. ejus, xiv, p. 482. Mémoires, cit.,viii, p. 415. Note, Vita	1121	Flies	Foigny near Laon
S. Bernhardi, i, No.58. Acta., SS. Aug. iv, p. 272	1121	Horseflies	Mayence
" " "	1320	Cockchafers	Avignon
Malleolus: De Exorcismis, Tract. ii,Mémoires, cit., viii, p. 411	14th cent.	Spanish flies	Mayence
Desnoyers: Recherches etc.	1460–1461	Weevils	Dijon
[Pg 318] Schilling: Chronica (Zurich MS), Hottinger: Hist. Eccles. Pars iv, pp. 317–21	1478	Inger (sort ofweevil)	Berne
Ruchat: Hist. Eccles. du Pays de Vaud	1479	Inger	"
Chasseneus: Consilia von Amira, p. 561	1481	Caterpillars	Macon
" " "	1488	Weevils	Beaujeu
Malleolus: De Exorcismis	"	Caterpillars	Coire
" " "	"	Beetles	Coire
Louandre: L'Épopée des Animaux	1500	Flies	Mayence
Mathieu: Hist. des Évêques de Langres, p. 188	1512–1513	Rats and Insects	Langres
Groslée: Ephémérides, 1811, ii, pp. 153, 168. Cf. Théophile Raynaud: Opusc, 1665, p. 482. Mémoires, cit., viii, p. 413, 418, 424	1516 (1506 according to some authorities)	Weevils	Troyes in Champagne

Sources of Information	*Dates*	*Animals*	*Places*
Habasque: Not. hist. sur le Litoral des Côtes-du-Nord, p. 89	1516	Locusts	Tréguier
Ménebréa: Jugements rendus contre les Animaux, p. 505. From Grenier: Documents relatifs à l'hist. du pays de Vaud	1536	Weevils	Lutry (on Lake Leman)
Bailly: Traité des Monitoires, ii	1541	Grasshoppers	Lombardy
Berriat-Saint-Prix in Thémis, i, p. 196	1543	Snails and Locusts	Grenoble
Ménebréa: Jugements rendus contre les Animaux, pp. 544, 545, 556. De Actis Scindicorum com. St. Jul., etc.	1545and1546	Weevils	St.-Jean-de-Maurienne
[Pg 323] Desnoyer, cited in Revue des questions historiques, v, p. 278. Von Armira, p. 567	1554	Insects	Langres
Lessona: I Nemici del Vino. Regist. Epir. Par. for May 8	1562	Weevils	Argenteuil
Chorier: Hist. du Dauphiné. Cf. Thémis, i, p. 196	"	Locusts	Valence
[Pg 325] Ménebréa: Jugements rendus contre les animaux, etc., pp. 546, 549	1587	Weevils	St.-Jean-de-Maurienne
Théophile Raynaud: De Monitoriis, p. 482. Mémoires, cit., viii, p. 429	16th cent. (first half)	Weevils and Grasshoppers	Cotentin
" " "	"	Weevils	Mâcon
Duboys: Hist. du Droit Crim. de la France	16th cent. second half	Weevils	Angers
Archives of Obwalden	"	Gadflies	Aargau
Leonardo Vairo: De Fascino. Cf. D'Addosio, cit., p. 115	"	Locusts	Naples

Sources of Information	*Dates*	*Animals*	*Places*
Lessona: I Nemici del Vino, 1890, p. 141	16th cent. second half	Weevils	Vercelli
Desnoyers: Recherches, etc., p. 13	1616	Rats and insects	Langres
[Pg 329] Marchisio Michele: Gatte ed. insetti nocivi, 1834, p. 63 *sqq.*	1633	Weevils	Strambino (Ivrea)
Crollolanza: Storia del Contado di Chiavenna, p. 455 *sqq.*	1659	Caterpillars	Chiavenna
Perrero: Gazzetta Litteraria di Torino, Feb. 24, 1883	1661	Weevils	Turin
Annales scientifiques de l'Auvergne, vii, p. 391	1670	Locusts	Clermont
Perrero: Gaz. Litter. di Torius, Feb. 24, 1883	"	Weevils	Turin
Dulaure: Description des principaux lieux de la France, 1789, v, p. 493 *sqq.* Mémoires, cit., viii, p. 412	1690	Locusts	Pont-de-Château in Auvergne
Agnel: Curiosités judiciaires et historiques du moyen-âge, p. 46. Cf. Manoel Bernardes: Nova Floresta ou Sylva devarios apophthegmas, etc. 5 tom. Lisboá, 1706–1747	1713	Termites	Piedade no Maranhãoin Brazil
Ménebréa: Jugements contre les animaux, p. 508	1731	Insects	Thonon
Desnoyers: Recherches, etc., p. 15	1826	Locusts	Clermont-Ferrand
Krauss, quoted by Amira, p. 573	1866	Locusts	Pozega in Slavonia
" " "	"	Grasshoppers	Vidovici in Slavonia
Desnoyer: Recherches, etc., p. 15	19th cent.	Locusts	Catalonia

boom-and-bust cycles related to environmental conditions. Thus, by the time a pest reached levels that prompted getting the Church involved, chances were that natural population control factors (disease, predators, parasites, lack of sufficient food) were already at work putting pressure on the pest population. Any actions of the Church at this point were likely to appear successful simply because they coincided with natural cycles. Therefore the appearance of success against a natural phenomenon that was both damaging and not understood prompted the actions of the Church to be credited with the decline of the pest.

THE PLAGUE AND ITS EFFECTS ON AGRICULTURE

The Black Death (1347–1351) was directly connected to insects through fleas, especially the oriental rat flea (*Xenopsylla cheopis*), as the transmitting agent. As blood-sucking parasites, there was no direct connection between fleas and agriculture. However, the indirect influences of the plague on European life, especially agriculture and diet, were significant.

Depending on the region, 30–80 percent of the population was lost (Alchon 2003). Even within a country, there was significant variation in mortality rates from one town or parish to another. Cohn (2008) states that more than 75 percent of the population of Florence died and that Italy in general had high mortality. Estimates of mortality for London are between 30 and 50 percent of the population. Germany and regions further north had somewhat lower mortality rates. Exact numbers were difficult to ascertain due to the paucity of good record keeping during that time (Zeigler 1969; Routt 2018). The large losses in human population affected numerous aspects of society, including agriculture; with the decline in population, there was a consequent decline in the demand for agricultural products. At the same time, the disease reduced the number of people available to work on the farms. These two factors resulted in large amounts of land being taken out of crop production (Thirsk 1997).

In an effort to use the available land with a reduced number of workers, much of it was converted to grazing land in place of cropland. Sheep in particular were raised for both wool and meat. Surprisingly, another market that rapidly expanded was rabbits. Rabbits, unlike hares, were not native and had been brought to England's Scilly Isles, Lundy Island, and the Isle of Wight in the late twelfth century. By the mid-thirteenth century, they were on the English mainland, but they were restricted to enclosed warrens and protected plots of land established by permit from the King (Veale 1957; Bailey 1988). Although rabbits were raised for food and fur prior to the plague, the market for their meat was small and targeted the upper class. However, as crop production became economically unfeasible, rabbit warrens grew in popularity to make use of otherwise idle land. As Bailey (1988: 12) states "… the drastic reduction in the human population after the mid-fourteenth century heralded a remarkable change in fortunes for commercial rabbiting." Rabbits required less labor than did crops, and provided furs as well as meat.

One consequence of the shortage of laborers was that workers began to demand higher wages from lords. They were able to do this as a result of the greater competition for their services. In many cases, the lords were willing to accommodate the demands of the workers, but this disruption in the status quo of societal relationships was seen as a threat by the upper class, including King Edward III. In 1351, shortly after the plague subsided, he issued the "Statute of Laborers," a decree that essentially made it a crime for workers to refuse to work for anyone wishing to hire them (The Statute of Laborers 1351, in Henderson 1896). In addition, workers who demanded higher wages could be punished. This law, however, was not very effective. Some employers used various "in-kind" methods to subsidize a worker's wages as a way of skirting the law. Others ignored the law outright and paid what was needed to be competitive. Dyer (1988) notes that as a result of the population shifts and the consequent restructuring of employer-worker relationships the plague produced, the economic conditions of the working class improved, and consequently, their diet improved as well, including their ability to afford rabbit meat. Archaeological evidence of this societal shift was found in the cookware used by the serfs. A significant increase in metal cookware over pottery had been observed to occur around this time period (Gunnell 2012).

The manorial system of agriculture was gradually eroded by the shifting socioeconomic circumstances. Peasants' leverage for bargaining increased as their ability to abandon their holdings and find work elsewhere increased. A vacant holding produced no income for the owner. Lords adjusted to the shifting balance by ceding workers increasingly more control over property and working conditions:

> The average size of the peasant holding grew in the late Middle Ages. Due to the peasants' generally improved standard of living, the century and a half following the *magna pestilencia* has been labeled a "golden age" in which the most successful peasant became a "yeoman" or "kulak" within the village community. Freed from labor service, holding a fixed copyhold lease, and enjoying greater disposable income, the peasant exploited his land exclusively for his personal benefit and often pursued leisure and some of the finer things in life. Consumption of meat by England's humbler social strata rose substantially after the Black Death, a shift in consumer tastes that reduced demand for grain and helped make viable the shift toward pastoralism in the countryside.
>
> (Routt 2018: 49)

Thus, the flea responsible for the Black Death in the 1300s played a significant role in changing societal socioeconomic patterns that, among other things, resulted in changes to the agricultural and dietary status quo that had been in place prior to the plague.

STORED PRODUCT PESTS

Global exploration by sailing ship greatly increased during the fifteenth and sixteenth centuries. The length of these voyages dictated the type of foods and materials that could be transported. Dry goods were less likely to be spoiled by mold, but were still subject to stored product pests. Likewise, live plants were not likely to survive a voyage, but seeds were frequently carried to start crops in new lands. As a result, accidental introduction of crop pests was not a significant issue during these early exploratory travels, but there was evidence for intercontinental transport of stored product species during this time period.

Early archaeological records indicated that many stored product pests originated in the Middle East or Far East and traveled with humans through migration and trade (King et al. 2014). Granary weevils (*Sitophilus granarius*), saw-toothed grain beetles (*Oryzaephilus surinamensis*), and flat-grain beetles (*Cryptolestes ferrugineus*) were common in pit-fill deposits, cesspits, and other archaeological sites in England and Germany that dated to the Renaissance period. In many sites, the granary weevils tended to be the most common. These insects were likely spread by the Romans throughout their empire.

Beyond the Roman Empire, stored product insects were found in Ireland and the Scandinavian region. Cesspits, shipwrecks, and other archaeological sites provided evidence of a variety of species from the twelfth through the sixteenth centuries (King et al. 2014). In addition, stored product pests such as granary weevils and saw-toothed grain beetles have been discovered in Iceland and Greenland from deposits dating to the mid-fourteenth century and earlier. Following their spread across Europe, these insects proliferated in the New World with European explorers. Jamestown, Virginia, established just after the early 1600s, contained evidence of numerous species; it was likely that the first ships to the Western Hemisphere introduced stored product pests.

The presence of some insect species was commonplace in the biscuits that voyagers lived on during their global exploration. The word "biscuit" traced back to the Latin words *bis coctus* which translates to "twice cooked." After the initial cooking, the biscuits were further dried in a slow-oven (Anon. biscuit). Biscuit gradually softens as it absorbs moisture from the air. In an effort to provide longer-lasting food for longer ship voyages, a four-step process was developed which produced very hard, dry biscuits called "hardtack". To avoid breaking teeth, hardtack was usually softened by dunking in coffee or cooking in some liquid.

Hardtack was a staple for these voyages (Heymsfeld n.d.) and was often found to contain "weevils" and "cockroaches." In reality, the "biscuit weevils" were larvae of what are now commonly called "drugstore beetles" (*Stegobium paniceum*). In low numbers, these small larvae, covered in biscuit dust, were likely eaten without the consumer's knowledge. At higher levels of infestation, sailors were known to bang the hardtack in an effort to knock the larvae from the biscuits before eating them. Highly infested biscuit might crumble when knocked on the table. Some sailors preferred to eat in the dark rather than see what they were actually eating. Larger grubs of the cadelle beetle (*Tenebroides mauritanicus*) were also commonly found in the hardtack supplies. These swarming grubs were called "bargemen" since they were often seen roaming around in the bread tray which was called a bread barge. Being predators, these grubs probably fed on the drugstore beetle larvae. The adults, being sizeable beetles, were incorrectly referred to as cockroaches (USS Constitution Museum 2014).

The steward's assistant on ships was often called "Dusty" or "Jack-the-dust," being associated with the dust from the hardtack as it degenerated, largely due to insect pests. To give an idea of the quantity of hardtack required, the records from relatively small expeditions ordered by Philip II, King of Spain, to raid the coast of England included 12,837 barrels of biscuit, with each barrel containing about 224 pounds. One record documented 41,400 pounds of biscuit having 2,420 pounds of dust. Magellan's journal from 1520 stated:

> Having in this time consumed all their Bisket and other Victuals, they fell into such necessitie that they were inforced to eate the powder that remayned thereof, being now full of wormes, and stinking like Pisse by reason of salt water.
>
> (Swinburne 1997: 313)

In a description of the cuisine of Columbus, mention was made of weevils in the hardtack and how the humidity on land in the Americas turned the hardtack into pulsing masses of weevils (Sorré n.d.). This observation was clear evidence that stored product pests from Europe were brought to the Americas and other parts of the world.

THE COLOMBIAN EXCHANGE—SCALE INSECTS, ANTS, AND AGRICULTURE

Just as agricultural labor in Europe was affected by the plague, insects also influenced agricultural practices in the New World. The "Columbian Exchange" was the beginning of trade routes connecting Europe, Africa, the Americas, and Asia. In 1492, Christopher Columbus landed in Hispaniola while seeking a route to China. After starting his first settlement, La Navidad, he returned to Spain with some of his crew to announce his progress and bring supplies back to those who remained in La Navidad. During his absence, the settlers at La Navidad struggled to survive and came into conflict with the local Taino natives in the process. Upon his return, Columbus found his first settlement in ruins, and in early 1494 decided to start a second settlement, La Isabela, elsewhere along the coast of Hispaniola. The Europeans had brought with them sugarcane (originally from New Guinea), coffee and bananas from Africa, and wheat from the Middle East. Just as importantly, they and future expeditions introduced a number of species that would significantly influence the ecology of the New World—some for the better, but most for the worse. These included mosquitos, cockroaches, rats, honeybees, dandelions, African grasses, and earthworms. These early introductions were only the beginning of an invasion that continued as ships transported goods from one continent to another (Mann 2011).

In 1516 the Spanish introduced African plantains to the New World by bringing a shipment to Hispaniola. E. O. Wilson (2005) speculated that these fruits were accompanied by one or more species of scale insects. It is likely that these insects easily spread to the bananas and other crops that by then were established on the island. It is also possible that their populations exploded as a result of a beneficial climate, abundant food, and especially, a lack of natural predators (Wilson 2005).

In 1518–1519, Bartolomé de las Casas, a missionary priest who lived in Hispaniola at that time, documented catastrophic destruction of Spanish orange and pomegranate orchards as well as cassia (Chinese cinnamon) plantations. He described thousands of acres that appeared "scorched and dried out." At the same time, he described the area as being overrun with stinging ants. It was so bad that colonists placed the legs of their beds in pans of water to keep the ants out of their beds so they could sleep. Ultimately, many of the colonists abandoned their homes and farms. Those who remained chose St. Saturninus as a potential savior to intercede for them. Following ceremonial processions and feasts to St. Saturninus, de las Casas reported that the plague of ants began to diminish (Mann 2011). Such proceedings were reminiscent of the insect trials taking place in Europe at the time.

The Spaniards attributed the destruction of the orchards to the highly visible and pugnacious fire ants. Wilson (2005) believed that populations of the ants, probably the tropical fire ant *Solenopsis geminata*, exploded as a result of their interactions with the scale insects. Ants and scale insects often form symbiotic relationships in which the latter provide sugary secretions for the former and receive protection in return. In the absence of natural predators, the booming populations of the scale insects could have caused tremendous crop damage as well as a concurrent explosion of the ant populations. This was probably the first record of ecological upset as a result of invasive species in the New World. If so, it was also one of the first examples of "ecological release" or the ability of an introduced species to thrive in an environment free of its natural predators, parasites, and diseases. Studies had shown that the tropical fire ant was native to the region, not introduced. However, the ants' ability to take advantage of introduced scale insects and of the ecological changes brought by agricultural conditions continued to cause problems.

Sugarcane, imported to the Caribbean, became an important cash crop due to Europe's demand for sugar. Until this time, honey was the only sweetener commonly available to Europeans (Berenbaum 1995). Both the French and the Spanish established large sugar plantations in the Caribbean islands in the sixteenth century, but ran into problems getting labor to work the fields. Europeans would not consent to work in the hot, buggy (including fire ants!) conditions. Natives were dying from the diseases brought by the settlers. Later, in the seventeenth century, English and French settlers on the east coast of what is now the United States also had trouble finding laborers to work the fields, largely because of disease. Africa, then, became the source of people who could work the fields under the harsh conditions and who were not very susceptible to the problem diseases. It was under these conditions that the slave trade grew and prospered (Albee 2014). Thus in another example of the interaction between insects, agriculture, and social issues, fire ants played a role in the African slave trade in the New World. As an additional "unintended consequence," Gullan and Cranston (2014) noted that new diseases were introduced to the New World through the slave trade. Malaria and yellow fever were the most significant of these. Thus, fire ants were at least partly responsible for agricultural problems, the slave trade, and the introduction of major diseases into the New World.

The agricultural impact of the tropical fire ants in the Caribbean was a severe blow to the early Spanish settlers in the region. Ironically, it appears that the Spaniards played a major role in the transport of these ants to become agricultural pests elsewhere. Gotzek et al. (2015) used DNA studies in an attempt to reconstruct the spread of the fire ant (*S. geminate*) throughout the world. DNA evidence points to a strong correlation between Spanish trade routes and establishment of tropical fire ants in new regions of the world (see Figure 3.2). The Spanish established the Spanish-Manila Galleon trade route that connected Acapulco, Mexico with Manila in the Philippines. Using Manila as a hub in Asia, regional trade routes connected Manila with China, Australia, India, and other places in the Orient. Likewise, the Spaniards connected to other locations in North, Meso-, and South America. Genetic analysis revealed a similar pattern with tropical fire ants, from southwest Mexico (near Acapulco) showing the more ancient lineage, ants near Manila showing genetic signs of having been derived from the southwest Mexican populations, and other Asian

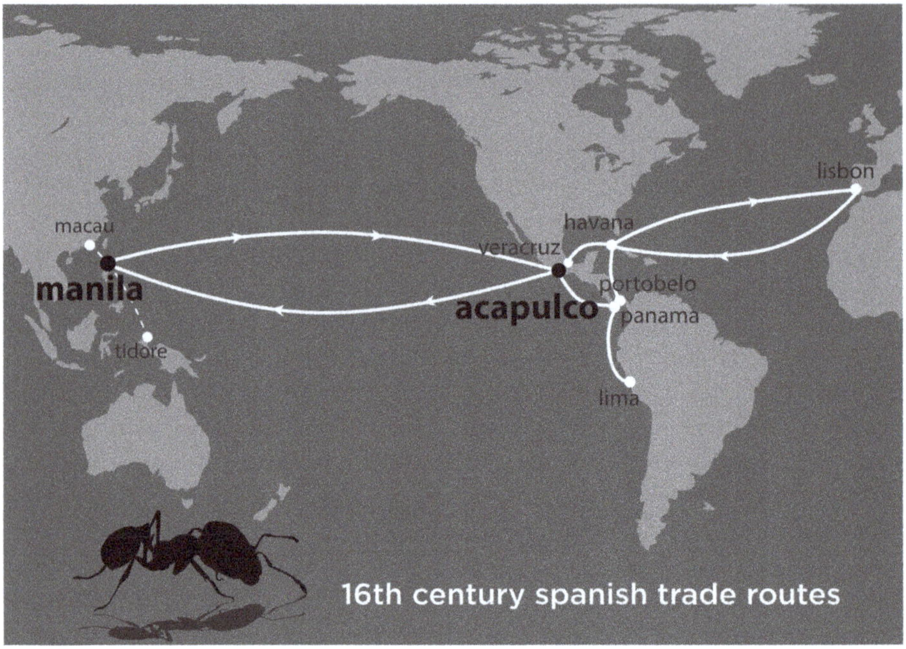

FIGURE 3.2 Map showing common trade routes of the Spanish, parallel to the spread of fire ants (*Solenopsis geminata*) from the Americas to other parts of the world. Illustration by Julie McMahon.

ant populations being derived from ants in the Manila area. The estimated age of the populations in the Orient dates to the early sixteenth century.

Gotzek et al. (2015) also cited numerous sources that established the tropical fire ant as a serious agricultural pest. Though ants' damage was most often through their support of honeydew-producing pests, they themselves had been documented as girdling citrus trees and deterring pollinators as well as being a nuisance to field workers. This strengthens Wilson's hypothesis that tropical fire ants and honeydew-producing insects were responsible for the sixteenth-century problems in Hispaniola.

ENTOMOPHAGY

Entomophagy, or eating insects, is a cultural activity that, where it exists, has probably been around for hundreds or thousands of years. Mitsuhashi (2016) includes 2,141 species of insects in the world that have been and/or are consumed by one or more peoples. Entomophagy occurs in 102 countries, and 699 species of insects have been recorded as food in the Americas

(Ramos-Elorduy et al. 2007). However, Europeans and their North American descendants do not look favorably on insects as food. Consumption of insects was probably normal in our primate ancestry and the loss of entomophagy is probably a function of "optimal foraging." In other words, it is not worth the time and effort to harvest insects if other sources of protein are readily available (Harris 1986; Evans et al. 2015). European attitudes towards entomophagy were fostered primarily during the Middle Ages when pork and other meats became so common in Europe that even the poor had access to meat (Smith 2013). At the same time, insects in Europe became associated with crop destruction and disease.

These regional attitudes came into conflict during the Renaissance when Europeans embarked on the age of exploration and conquests. They considered the eating of most animal foods with which they were not familiar with as exemplifying behaviors typical of primitive and dirty peoples. Pagden (1986) stated that because the Indians could not distinguish between things that were "edible" and "inedible," they were considered barbarians. Anecdotal confirmation of these attitudes came from the explorers themselves. For example, after the Aztecs had been introduced to the European diet, Francisco López de Gómara in 1552 commented that "... what is all the more amazing is that they eat such bugs and dirty animals when they have good bread and wine, fruit, fish and meat" (Lopez de Gómara [1552] 2003: 93).

Early evidence of entomophagy was provided by Bernardino de Sahagún, a sixteenth-century Spanish missionary in Mexico. He learned Nahuatl, the language of the Aztecs, and used that knowledge to do extensive ethnographic studies of the Native Americans in Mexico and the Central American region. He arrived in "New Spain" in 1529 and spent sixty-one years living with and studying the Nahuat people. In his monumental work *The General History of the Things of New Spain*, also known as the *Florentine Codex*, he recorded ninety-six species of insects as part of the diet of the Aztecs. It was not always easy to tell whether a particular animal being referred to was an insect, as many unrelated creatures, including insect larvae, were called worms. In addition, taxonomy was still in its early stages at that time so names and relationships of insects were unclear by today's standards. Yet almost all the insect and insect-like creatures mentioned by Sahagún include a phrase such as "Cómenlos" meaning "They eat them." In addition to insects, Sahagún wrote about other animals, plants, the society, gods, and more. During his time with the Nahuat, he was able to teach many of them, including some of the royalty, Spanish and Latin. Thus, he collected stories, terms, and customs from a variety of individuals and was able to cross check his information. He used the Nahuat to help him compile, write and illustrate his twelve volumes. Yet even with his anthropological approach, he too lectured the Indians on their diet and told them they could become pure and wise like the Castilians if they ate better food (Earle 2010).

Ramos-Elorduy and Pino Moreno (1989a) described how insects contributed to the diet of Mexicans before the Spanish Conquest. They listed 247 species consumed in Mexico, of which seventy are described. In doing so, they incorporated some of Sahagun's accounts from the early sixteenth century. Some of the drawings from the *Florentine Codex* were added to help establish the consumption of these insects in the past. Some insects such as escamole (larvae of the ant *Liometopum apiculatum*) and ahuauhtli (eggs of Corixidae, aquatic Hemiptera), were of such value in the Aztec culture that they were used to pay taxes to the Emperor Montezuma. Sahagún stated that the Emperor Montezuma and the Aztec kings who preceded him, when celebrating the god Xiuhte-cutli, had ahuauhtli prepared and brought into Tenochtitlan by native runners from Texcoco so that the Emperors would have them fresh for breakfast.

Another prominent contributor to the documentation of insect-eating behavior of early Mexicans was Francisco Hernández de Toledo, a Spanish naturalist and physician. He spent seven years (1570–1577) exploring and cataloging the plants and animals of "Nueva España." The following is a translation of his description of the abundant corixids (bugs with the common name of water boatmen but referred to here as a "fly") and their eggs that were harvested from Lake Texcoco (Parsons 2010: 125) which has since been drained and is now the site of Mexico City:

> The axaxayácatl is a small, lacustrine fly ... which in certain seasons is collected with nets from the lake in such great quantities that great numbers of them are cut up and mixed together to form little balls, which are sold in the markets throughout the year; the indians cook them in salty water wrapped up in maize husks, and prepared in this way they comprise a good food, abundant and agreeable ... There are two species of these flies, one large and the other smaller.
>
> (Hernández 1959: 390)

> A great quantity of a certain substance called ahuauhtli, with a fishy taste, is taken from the Mexican lake [Lake Texcoco]. It looks like a poppy seed, and it is the eggs of the axacayácatl ... It is gathered by throwing into the lake, where the waters are most turbulent, loosely twisted cables as thick as a man's arm or thigh. The [eggs] shaken and swirled, adhere to these, from which the fishermen remove them and store them in large vessels. They make tortillas [from it] similar to corn ones, or the balls they call tamales ... or they save it, split into portions and wrapped in corn husks, toasting or cooking it at a later time.
>
> (Hernández 1959: 393)

Ahuauhtli appears to have been an exception for the Spaniards in their assessment of what was considered "inedible" foods of the Aztecs. García (2002: 191) noted that at the Friday markets "the Spanish favorites were fish, frogs, duck and, what can be surprising, the eggs of axayácatl[1] insect, called ahuauhtli, which were prepared during fasting."

Mesoamerica had no pasture animals to provide protein from a landscape that was not highly productive from an agricultural perspective. The protein sources provided by the saline Lake Texcoco were key in allowing the Aztec population to flourish in pre-Columbian times (Parsons 2005). Not only did the aquatic insects, fish, and plants provide food directly, but they attracted large numbers of overwintering birds which were also harvested by the Aztecs. Estimates of harvested insect products from the lake are in the range of thousands of metric tons. Combined with their nutritive value (Bodenheimer 1951; Ramos-Elorduy et al. 1984), these insects were a major source of protein for the Aztecs.

However, not everyone lived near the water. Below are described some of the insects used as food by the Aztecs and Mayans in dryer habitats. All of these can still be found for sale in markets today.

Chapulines (grasshoppers) can be found for sale at many markets and street corners in Mexico, especially in the southern state of Oaxaca. The word itself is Nahuat in origin which is not surprising given that chapulines were a rich source of protein and fat for the Aztecs. Sahagún recorded chapulines as a common food among the Aztecs. Hernández found the same, expressing the following: "... these Indian Westerners eat tadpoles willingly, which our countrymen are horrified to see and even to name, and not despise fried grasshoppers and ants ..." (Hernández 2007: 21).

Escamoles have now become an expensive meal at restaurants in Mexico City. However, they were a welcome springtime treat to the Aztecs. The name was derived from the Nahuatl term "azcamoli" which itself came from azcatl (ant) and molli (sauce or puree). Though commonly referred to as ant eggs, these are the mature larvae of the queen ants of *Liometopum apiculatum* or velvety tree ant and related species. Nests are usually found in the earth under agave plants and are excavated until the larder of ant larvae and pupae is reached. These ants do not sting, but are vicious biters and it takes fortitude, and a taste for escamoles, to endure long enough to harvest the larvae. *Zompopos de Mayo* are the large queens of the leafcutter ants. The winged queens emerge in large numbers in May, usually after rains soften the ground. They are filled with eggs and easy to collect. The abdomens are removed and eaten like peanuts, whose flavor they are said to resemble (Ramos-Elorduy and Pino Moreno 1989b).

Indians obtained honey from a variety of insect sources (Curran 1937). The *nequazcatl* or honey ants have a caste that produce honey stored in their bloated abdomens. These "honeypot" ants serve to feed the other ants in the colony. The ants were (and are) eaten alive by biting off the abdomen or just eating the entire ant. (The use of stingless bees to provide honey is discussed below.)

Garcia-Herrera (2010) wrote about the many uses of plants of the genus *Agave* in Mexico, one of the most prominent being food. Ironically, in the list of foods that come from the agave plants are three insect species—*gusanos blancos* (white worms), *gusanos rojos* (red worms) or *chinicuiles,* and *picudo del maguey* (agave weevil larvae). While agaves are harvested and used for food, fiber, roofing, and other things, these insects are also harvested and used as food. It is not uncommon that the insect "pests" are more coveted and valuable than the plants they infest. The *gusanos blancos* are the large larvae of the tequila giant-skipper butterfly (*Aegiale hesperiaris*). These were and are highly prized for their flavor. The second caterpillars are smaller and red. These *gusanos rojos* are found primarily in the roots of the agave plants, often in good numbers, and are the larvae of a carpenter moth, *Comadia redtenbacheri* (Curran 1937). These are the same larvae that can be found in the bottles of some brands of mescal. The last species (*Scyphophorus acupunctatus*) is less well known, but still prized. They grow only in the agave that has been already used for the production of pulque, a drink made by cutting the flowering stalk of agave and harvesting the "honey water" that accumulates in the resulting cavity (Ramos-Elorduy et al. 2007).

While we have a significant knowledge of the insects eaten by the Aztecs, Gene DeFoliart (2002) collected extensive worldwide information on entomophagy which he published on a University of Wisconsin website; unfortunately, he died before putting it into printed form. Here we present some of his compilation; we assume that current habits of aboriginal/native peoples were used during the Renaissance period as well.

Australia: Aborigines regularly eat a variety of beetle and lepidopteran larvae including witchetty grubs (*Xyleutes leuchomochla*) and Bogong moths (*Agrotus infusa*). Honey ants are also favorites.

Japan: In both historical and modern Japan, the most popular and widely eaten edible insect has been the rice-field grasshopper (*Oxya japonica*) which, fried and slightly seasoned with soy sauce, is known as inago. Silkworm pupae and brood from bees and wasps are also highly favored.

Philippines: DeFoliart quoted a newspaper article by Domoguen as saying "Eating insects is an old custom among the various minority tribes in the highlands of North Luzon. The habit is due to food needs and a way of reducing the pests which attack food crops, according to the entomologists."

North American Indians: In addition to grasshoppers and locusts, shore flies on Mono Lake and other alkaline western lakes were highly prized and harvested in abundance in season. Brewer in chapter four of his journal (Brewer and Farquhar [1930] 2003) observed that:

> The number and quantity of these worms and flies is absolutely incredible. They drift up in heaps along the shore – hundreds of bushels could be collected. They only grow at certain seasons of the year. The Indians come far and near to gather them. The worms are dried in the sun, the shell rubbed off, when a yellowish kernel remains, like a small yellow grain of rice. This is oily, very nutritious, and not unpleasant to the taste, and under the name of koo-chah-bee forms a very important article of food. The Indians gave me some; it does not taste bad, and if one were ignorant of its origin, it would make fine soup. Gulls, ducks, snipe, frogs, and Indians fatten on it.

Ecuador: Barragán (2009) documented observations from some of the early writers about insects in Ecuador. While a number of insects are included in their native diet, two beetles are particularly important. A scarab beetle (*Platyicoelia lutescens*) called "catzo blanco" is used in a seasonal fall dish in the Quito area, while the weevil larva known as "chontacuro" is found in markets in certain regions of the Amazon basin.

Libya: Bodenheimer (1951) cited the discussion by the Muslim zoologist ad Damiri of the consumption of locusts according to Islamic tradition which was published at the dawn of the fifteenth century. Leo Africanus (1896) noted that in spite of the destruction caused by locusts, peoples in the Arabian desert region and Libya welcomed the coming of the locusts. They would dry them in the sun and beat them into a flour.

MELIPONICULTURE: BEEKEEPING OF STINGLESS BEES

The cultivation of stingless bees was present in Mesoamerica long before Europeans introduced honeybees. The oldest evidence dates from the late preclassic period (between 300 BCE and 200 CE). The relationship of the native Mesoamericans with bees was complex as the latter were considered sacred. There was a god of the bees, Amucen kab, and ceremonies were held in order to guarantee the protection of the bees as well as the production of honey (Thompson 1990).

In the sixteenth century, Friar Diego de Landa, the Bishop of Yucatan, observed and recorded the lifestyles of the Mayans that lived in the area. He arrived in 1549 and wrote "Relación de las cosas de Yucatán" around 1566. The original copies were lost. The current version surfaced around 1660 and an English translation entitled "Yucatan before and after the conquest" (1937) is referenced here. De Landa commented on two kinds of stingless bees, smaller than the European species, used by the Mayans as sources of honey. The bee most associated with the Mayans was *Melipona beecheii*, known as xuna'an kab or colel-kab ("royal lady") in the Maya language. While these bees were wild, the Mayans cultivated them using hollow wooden tubes with stone or hardened mud caps. The cap on one end had an entrance hole. These tubes were stored and honey was periodically harvested from them. These bees do not build combs, but store the honey in pots of wax. Mayans would use a stick to break up the pots and collect the honey that poured out. De Landa noted that these bees are harmless, even when honey is harvested. The honey, wax, and royal jelly were all used for food, medicines, and ceremony (Hirst 2017). The *Codex Mendoza* recorded the use of wild bee honey in the preparation of chocolate, along with flowers, vanilla and chile (Berdan and Anawalt 1997). Diego de Landa (1937) mentions a wine made out of honey and the bark of the lilac tree (*Lonchocarpus violaceus*) to make a honey mead called balché. This intoxicating drink was and still is used in ceremonies; the drink was prohibited in the sixteenth century by the Inquisition.

While Friar de Landa was one of the first to describe raising stingless bees for the Europeans, the *Madrid Codex*, a Mayan text believed to be from the thirteenth to fifteenth centuries, contains a detailed description of the practice of beekeeping by the Maya people. Likewise, the *Codex Mendoza* is a text combining Aztec pictograms with Spanish commentary. Written in the mid-sixteenth century, after the Spanish Conquest, it records Aztec provinces and enumerates the tributes (taxes) paid by each town (Hogue 1993; Bianco et al. 2017). The tributes depended on the resources of the particular regions. Many of the southern provinces, including regions presently in the Mexican states of Guerrero, Puebla, and Oaxaca, included bee honey as part of their tribute.

Beeswax and honey were also used to trade for cacao and precious stones, and under Spanish rule, the Mayans citizens were taxed as well. Records from 1549 show 150 villages paying a combined tax of three metric tons of honey and 281 metric tons of wax. Francisco Hernández de Toledo (1959) also recorded the harvesting of honey, and noted that the natives relished the bee larvae as well.

The introduction of European honeybees is generally associated with the English colonists who brought them to Jamestown in 1622. Brand (1988) provided several pieces of indirect evidence that European honeybees might actually have been brought over by the Spanish in the early sixteenth century and introduced to Cuba or other Caribbean islands. That argument has yet to be fully resolved. In any case, the introduction of *Apis mellifera*, the invasion by African bees in the late twentieth century, the loss of forest, and climate change-induced storms have all contributed to the decline in native stingless bees. In the case of *Melipona beecheii*, a loss of the Mayan cultural traditions and knowledge are also contributing to a drastic decline in working hives (Smithsonian 2005).

CONCLUSION

Despite the dearth of scientific studies on insects and food during the period 1300–1600, the contributions of anecdotal evidence and observations on the topic during the period cannot be ignored. The reports and journals of explorers, colonists, and missionaries, along with church records, religious writings, and the field studies of a few naturalists have provided valuable information on insect pests and pest control, the regional and global spread of insects and insect trials. More recent archaeological studies and meta-analysis of period writings has confirmed many of these previous ideas and provided an expanded perspective on the many connections between insects and food. From this, we can see the significant influence insects had on food production and storage, especially as global interconnectedness advanced during the Renaissance. This influence varied from direct consumption of food intended for humans, to effects of disease on agricultural practices and diet. Finally, the importance of insects as food or as providers of food through their by-products varied widely by region and culture, resulting in opportunities for expanding perspectives as well as reinforcing cultural bias.

CHAPTER FOUR

Insect Products

SHERILYN G.F. SMITH AND GENE KRITSKY

The major insect products of the Medieval Age, honey, beeswax, silk, ink and red dyes, continued to be valuable commodities at the start of the Renaissance, and the uses and technologies associated with these products expanded during the period. The discovery of the New World and the Colombian Exchange led to the introduction of new insect products to Europe and to the Americas. Another major change during the Renaissance was our understanding of how these insect products were procured and utilized. The invention of the movable type printing press ushered an expansion in books that could be produced in greater number, which reduced their prices and made them widely accessible. Moreover, many of these books dealt with topics outside of religion, which predominated in the illuminated manuscripts of the Medieval Age. Books and treatises were written on beekeeping, sericulture (culture of silkworms), silk fiber production and textile manufacture, the making of dyes, inks, and medicines, and reports describing the plants and animals of the New World and the practices of the Indigenous peoples there. Insects and insect products figured prominently in these works.

Some of these early printed books also utilized woodcut illustrations which further added to their value. Woodcut illustrations originated during antiquity in China, and eventually made their way to Europe from China along with the process of making paper. Woodblock prints were made by the illustrator drawing the image on a block of wood and carefully cutting away the wood so that the cut away areas would not be inked. The images permitted printed books to include flourishes that were reminiscent of similar designs in illuminated manuscripts, and other images illustrated, sometimes in great detail, topics discussed in those books, including the various types of beehives and beekeeping practices employed during the Renaissance.

BEES AND BEEKEEPING

One of the earliest beekeeping images dates to 1499, and it depicts two upright log hives with bees flying about them (Figure 4.1). The hives were furnished with an entrance cut at the base and were covered with round wooden tops. A woodblock illustration from 1560 shows three sets of two log hives stacked up one another. It is not clear from the illustration if the hives were "supered," meaning the set accommodated one large colony, or if the hives were closed off from each other. Supering with box hives was practiced in 1595 (Forster 1975; Kritsky 2018).

Both wicker and straw hives were illustrated with woodblock illustrations (Figures 4.2 and 4.3). In many of the cases the hives were placed upon benches that were protected from the elements by a slanted roof that would permit rain water to be shed behind the hives whereas other illustrations have the roof overhanging the hives, but shedding the water off in front of the hives (Forster 1975; Crane 1999).

It is a common cliché that a picture is worth a thousand words, but the major impact of the printing revolution was the publication of beekeeping books written by beekeepers. The bestiaries and other books produced during

FIGURE 4.1 Two upright log hives. Prüss, Johann. 1499. *Hortus Sanitatis*. Straßburg. Public domain.

FIGURE 4.2 A beekeeper with two wicker skep hives. Münster, S. 1546. *Cosmographia*. Basil. Public domain.

FIGURE 4.3 A beekeeper ready to harvest honey from two straw skep hives. Crescentiis, P. v. 1518. *Opus Ruralium commodorum*. Straßburg. Public domain.

medieval times were often fanciful in their narratives about bees. However, by the 1500s readers were able to learn from successful beekeepers by reading their books. The earliest book in English that included some beekeeping information was by John Fitzherbert and was published in 1523. The book, titled *Book of Husbandry*, only had one section on beekeeping:

> **122. Of bees.**
> Of bees is lyttell charge but good attendaunce; at the tyme that they shall cast the swarme, it is convenient, that the hyve be set in a garden, or an orchyarde, where as they maye be kepte from the northe wynde, and the mouthe of the hyve towarde the sonne. And in June and July, they do most commonlye caste, and they wolde have some lowe trees nyghe unto them before the hyve that the swarme maye lyght upon, and whan the swarme is

knytte, take a hyve, and splente it within with thre or foure splentes, that the Bees maye knytte theyr combes therto, and annoynte the splentes, and the sydes of the hyve, with a lyttell honye. And if thou have no honye, take swete creame, and than set a stole or a forme nyghe unto the swarme, and laye a clene washen shete uppon the stole, and thanne holde the smalle ende of the hyve downewarde, and shake the Bees in to the hyve, and shortely sette it uppon the stole, and turne uppe the corners of the shete over the hyve, and to leve one place open, that the Bees may come in and out: but thou mayst not fight nor stryve with theym for noo cause, and to laye nettyls uppon the bowes, where as they were knytte, to dryve them from that place, and soo watche them all that daye, that they go not away, and at nyght, whan al be goone up into the hyve, take it away and set it where it shall stande, and take awaye thy shete, and have claye tempered to laye aboute it uppon the borde or stone, where it shall stande, that noo wynde comme in, but the borde is better and warmer. And to leave an hole open on the south syde, of three inches brode, and an inche of heyghte, for the Bees to come in and out. And than to make a coverynge of wheate strawe or rye strawe, to cover and house the hyve about, and set the hyve two fote or more from the erthe upon stakes, soo that a mouse can not come to it, and also neyther beastes nor swyne. And if a swarme be caste late in the yere, they wolde be fedde with honnye in wynter, and layde uppon a thynne narowe borde, or a thynne sclatte or leade, put it into the hyve, and an other thynne borde wolde be set before every hyves mouthe, that no winde come in, and to have foure or fyve lyttell nyckes made on the nether syde, that a bee maye comme out, or go in, and so fastened, that the wynde blowe it not downe, and to take it up whan he wyll. And that hyve, that is fedde, to stoppe the mouthe cleane, that other Bees come not in, for if they doo, they wyll fyghte, and kyll eche other. And beware, that noo waspes come in to the hyve, for they wyll kyl the Bees, and eate the honny. And also there is a bee called a drone, and she is greatter than an other bee, and they wyll eate the honny, and gather nothynge: and therfore they wolde be kylde, and it is a sayenge, that she hath loste her stynge, and than she wyl not labour as the other do.

(Fitzherbert 1523: Ch. 122)

Fitzherbert's small entry covered points that could not be not gleaned from a woodcut. He discussed where to place the hives, when to be on the lookout for swarms, how to hive a swarm, and how to protect the hive with a straw hackle covering. He also discussed how to protect the hive against pests such as mice, swine, and wasps.

Pests of honey bees were noted in other books as well. Konrad von Megenberg's *Das Buch der Natur*, based on Thomas de Cantimpré's thirteenth-century *Liber de natura rerum*, was first published in 1475 and included an illustration of skeps on a low table and protected by a lean-to. Running about under the table were mice (Crane 1999). The Library of Congress copy of the 1481 edition of *Das Buch der Natur* includes a different illustration of invertebrates along with two skeps on a small table to protect the hive from snails and ants (von Megenberg 1481; Library of Congress).

The first English book dedicated entirely to beekeeping, titled *A Treatise concerning the right use and ordering of Bees,* appeared in 1593, authored by Edmond Southerne. We have few details about Southerne other than that he was a "gentleman" who lived during the reign of Queen Elizabeth I. Southerne wanted to encourage the keeping of bees for the economic benefits and to lessen the fear of being stung (Kritsky 2010). Southerne's writing style is more straightforward compared to earlier beekeeping writings. His section on choosing and preparing a hive is a typical example:

> To choose your Hives.
> The best time is to have your Hives made about Christmas, for then you may have them better cheape, and straw is best in season, and the briers that they bind them withall are then strongest, and will indure the longer, but in any wise let not your Hives be above fifteen or sixteen rolls at the most, both of great and small, but if your Hives be made of twigs, as in some countries, the use, so they contain not above half a bushel apiece, it is enough.
>
> To prove that great Hives cannot be profitable.
>
> You are to note that when the Hive is bigge, it will aske long time and great labour for the Bees to fill, so the first year they cannot have time enough to doe it as they may the little Hives: and the next year when they should breed Bees, they are busied about filling up their Hives: for until the Hives be full of Combes, they will not swarme, and then the Spring being farre spent before they goe to breeding, the swarmes must needes come late, which doth both endanger the swarme, and also the olde stocke, for that the olde stock will have no pleasure to gather honey till the swarme be gone forth, which being late, the freshest flowers must needes decay before the Bees begin to worke, whereof proceedith many times such death of Bees.
>
> The manner how to dresse Hives before you put in Bees.
>
> First after you have bought your Hive, you must with a little knife proune and cut away all the little superfluous tickling strawes within side the Hive, and let it be a smooth as possible: for if those stawes should be let alone, when the Bees are new put into the Hive, they would scares like of it, for that they cannot abide such strawes, no more than haires or any other ragged

> thing: and if by chance they doe tarrie in such a Hive undressed, there they will of themselves be so long in their manner in scratching and biting away such paltrie, that they might have filled half the Hive with wax and honey in that space, if it were to be done for them. When you have so trimmed your Hive, then take Sallow, Willow, or Hazell stickes, which being cleft in the middle and cleane shaven, you may put six parcels into a Hive, that is to say, t'was crossed foure breadth within side, and the other four foure parcels from the top to te mouth, being stickes fast between the squares of the crosse stickes, is the best way you can spray your Hive, both for ease for the Bees, and also for staying the combes within side: and then, Fennel, Oak, Elme, or Ashen leaves being dipped in faire water with a little honey, besprinkle the Hive, or if you want honey, a little milke, if you have no milke, faire water will serve for a shift, and when you have so done put in your Bees, and they wil tarrie: but if you have no new Hives readie, then you may take an old Hive and use that as aforesaid. But if your Bees be so froward that they will not tarrie in any Hive, then take your Hive whether it be old or new, and pull out the stickes, and put therein two handfuls of Barley or Pease, but Mault is the best if you have it, and let a Hogge, a Pigge, or a Sow eate it, turning the Hive with your hands as he eateth, that the froth that he maketh in eating, may remaine in the Hive, then wipe the Hive againe with an old woolen or linen cloath, then thrust in your stickes as aforesayd, and so the Bees being put in they will abide without further trouble: but this is specially to be done to an old Hive rather than a newe, and you may doe it a day or two before you occupie it, and it will serve you will enough.
>
> (Southerne 1593)

The quality of Southerne's information is remarkable. Unlike other treatments, he delineates the size differences between wicker and straw hives, and he describes in detail how to prepare skeps for the bees. The hives should be placed on a stand about two foot off the ground to reduce the effects of wind, and he specifies that the beestone or bottom board that serves as a landing platform for the bees must be two hands wider than the hive to give bees ample space to "alight." Moreover, the hive should not stand too tall on the beestone or board to reduce the likelihood that snails and mice might enter the hive.

Southerne's book is focused on skep beekeeping, which required the death of the bees in order to harvest the honey. At this time some beekeepers were "driving" their bees out the hive in an effort to give the bees a new hive when the beekeeper harvested the honey in midsummer. Southerne did not recommend that practice because it killed too many larval bees and was generally not very successful. He did write that if you have determined to kill a stock of bees, that "you may let it swarm a second time during the late summer before killing the Bees."

A fourteenth-century illustration shows a beekeeper cutting out honeycomb with a knife and placing it in a bowl. No bees are illustrated on or around the hive suggesting that they were killed prior to the harvesting of the honey, as was a common practice. An Italian manuscript in the British Library dating to 1485 shows beekeepers working hives next to a large fire pit, presumably to kill the bees (Kritsky 2017).

The common method of killing bees was to dig a pit and start a brimstone fire. The skep would then be held over the fire and the shaken into the pit. This had the unfortunate effect of occasionally adding a foul odor to the honey. However, during the late Renaissance another method was becoming common, as described by Gerarde in 1597:

> ***Fungi lethales, [...] non esculenti.*** **Poysonous Mushrums.**
> There is also found another, set forth under the title *Fungus virilis penis arecti forma,* which wee English, Pricke Mushrum, taken from his forme.
> *Fungus orbicularis,* or *Lupi crepitus,* some do call it *Lucernarum fungus:* in English, Fusse balls, Pucke Fusse, and Bulfists, with which in some places of England they use to kill or smolder their Bees, when they would drive the Hives, and bereave the poore Bees of their meat, houses, and lives: these are also used in some places where neighbours dwell far asunder, to carry and reserve fire from place to place, whereof it tooke the name, *Lucernarum Fungus:* in forme they are very round, sticking and cleaving unto the ground, without any stalks or stems; at the first white, but afterward of a duskish colour, having no hole or breach in them, whereby a man may see into them, which being troden upon do breath forth a most thin and fine pouder, like unto smoke, very noisome and hurtfull unto the eies, causing a kinde of blindnesse, which is called Poor-blinde, or Sand-blinde.
>
> (Gerarde 1597: 1385)

The smoke from the puffball was quite effective at killing the bees without changing the flavor of the honey. It was often used with a bellows that would blow the smoke into the hive, which would be enclosed within a bag (Gerarde 1597; Crane 1999; Kritsky 2010).

The Renaissance was the first time that we see protective clothing being used while working the bees. Medieval illustrations show the beekeepers working the hives and harvesting honey without any type of protective clothing. The earliest illustration of such protection dates between 1320 and 1330 and shows a beekeeper with a cloth draped over his or her head. Fitted cowls that were placed over the beekeeper's head and held in place by a hat were in use by 1500 (Kritsky 2017). During the sixteenth century, flat screens made of plant fibers were fitted in the cowl to provide the beekeeper with a better view. Peter

FIGURE 4.4 A beekeeper carrying a tall, finely made straw skep hive. Ink illustration by Pieter Bruegel the Elder in 1568, https://www.wikiart.org/en/pieter-bruegel-the-elder/the-beekeepers-and-the-birdnester-1568. Public domain.

Bruegel the Elder's illustration of beekeepers show them wearing full suits with a screened hood (Crane 1999) (Figure 4.4).

In all of the cases where protective clothing is used, the beekeeper is illustrated striking a small drum. This practice, known as tanging, was thought to encourage a swarm of bees to settle down on a nearby branch, permitting the beekeeper to capture it in a bag so it could be placed in an empty hive (Crane 1999; Kritsky 2017). Southerne (1593) did not think much of tanging. He wrote:

> When the swarme is up it is not good to ring them, as some doe, nay it is a common thing where there is no experience, to keep a stirre and lay upon either a Bason, kettle or frying pan, taking (as the common proverb is) great paines, and have little thanks; for by such meanes they make the Bees angie, and goe further to settle, the otherwise they would, or els creepe close to the bodie of a tree, which must meeds be troublesome, for that they cannot abide such noyse.

Tanging was also used to alert neighbors that a swarm had issued from a hive, and to whom the swarmed belonged (Southerne 1593).

Many beekeeping books organized beekeeping activities around the calendar. Southerne made seasonal recommendations for various tasks. For example, new hives were built around Christmas, hives should be cleaned of dead bees that would have accumulated under the skep around the Ides of March, and inspecting the bees to determine which hives should be harvested should occur at Bartholmewtide or August 24. These practices became known as bee calendars and were commonly included in books for cottagers (Southerne 1593).

Thomas Tusser's *A Hundreth Good Points of Husbandrie* was first published in 1557 and enlarged to *Five Hundred Points of Good Husbandrie* in 1573. In keeping with many books at the time, Tusser's advice was arranged as verses to help the beekeepers remember what tasks should take place. The following is from Tusser's 1580 edition:

September
Drive Hive good conie
For wax and honie
No driving of Hive
Till year's past five
Good dwelling five bee
Or hence goes shee

Now burn up the Bees, that ye mind for to Hive,
at midsummer drive them, and save them alive:
Place Hive in good air, set southly and warm,
and take in due season, wax, honey, and swarm.
Set Hive on a plank (not too low by the ground)
where herb and flowers, may compass it round:
and boards to defend it, from north and northeast,
from showers and rubbish, from vermin and beast.

December
Helpe Bees, sweet conie
With licour and honie
Go look to thy Bees, if the Hive be too light,
set water and honey, with rosemary dight;
which set in a dish, full of sticks in the Hive,
from danger of famine, ye save them alive."

May
Watch Bees in May,
for swarming await,
Both now and in June,
mark master Bees tune.

Take heed to thy Bees, that are ready to swarm,
the loss thereof now, is a crowns worth of harm:
let skillful be ready, and diligence seen,
lest being too careless, thou loosest thy beene.

(Tusser 1580)

Sweet conie was a term of endearment for the woman of the house, who often took care of bees. Women's role in taking care of bees would continue to grow during the Age of Enlightenment, and eventually books would be written just for women beekeepers.

Many of the written records during the Renaissance focused on beekeeping in central and southern Europe. One text, however, opened up a window on beekeeping in Sweden. Olaus Magnus, originally Olof Månsson, was born in 1490 in Skånnige. His brother Johannes was Sweden's last Catholic Archbishop, and after Johannes's death in 1544, Magnus was appointed as his successor. The title meant little as Sweden was no longer taking orders from the Vatican, thus Magnus remained in Rome until his death in 1557. Magnus's major contribution was his 1555 work, *Historia de Gentibus Septentrionalibus* or *Description of the Northern Peoples*. The work included the first details of how beekeeping was practiced in Sweden. It was also peppered with woodcut illustrations that amplify the sometimes fanciful text. The tome is organized as twenty-two books subdivided into chapters; Book 22, Chapter 9 starts with "The territories of the North are rich in great varieties of products, necessities and luxuries of alike, which may be shared out for their own and the rest of the world's use. So too, by the remarkable bounty of provident Nature, they have a copious yield of honey" (Magnus 1996, 1998b).

Unlike Southerne's book, which is based on firsthand experience, Magnus uses writings by Pliny, Aristotle, and others to fill in details of management. What is useful for our examination are the observations that Magnus was thought to have remembered. Magnus, like so many others, promoted tanging. He wrote in Chapter 10, "To manage [hiving swarms] more easily, you drive them out by smoke from the caves or hollow trees where they so often cluster, and herd them together by clapping your hands or jangling brass implement" (Magnus 1998a).

One new observation involves migratory beekeeping, moving the bees during the year to get better pasturage. In Chapter 11 Magnus states:

> [Bees] are normally carried to distant places on your back, with their hives closed up, but where the terrain is flat and marshy, it is better to transport them on carts. Fifteen to twenty hives can be comfortably lodged in a single cart, as they do in Brabant and Flanders when they take their hives to level pasture at particular times in summer, to give them better nourishment.

An accompanying woodcut illustrates this practice showing a hooded beekeeper tanging with a bell in the distance to hive a swarm in a straw skep. The hives are placed on a table protected by a roof. In the foreground is a cart being pulled by two horses that is transporting sixteen or more skeps (Figure 4.5).

FIGURE 4.5 Skep hives being transported on a wagon. Magnus, O. 1555. *Historia de gentibus septentrionalibus*. Rome. Public domain.

An unusual practice that Magnus describes in Chapter 17 is how the bees can be protected during the harsh northern winters:

> The men of these northern lands have quite a clever device for protecting [their hives] through the winter months and keeping them from sickness or decrease in numbers caused by the biting frost. In mid-September they cover the hives completely with rye-straw or reeds and then, when the first cold spell arrives, they spill water over them and let it freeze. Then they shovel snow up high on top of the straw like a long envelope, so that however hard a frost eventually sets in, the bees cannot be killed.
>
> (Magnus 1998b)

Other bees would be housed in split logs that were covered with pine bark and placed in thick snow for up to eight months.

The honey was harvested during the summer in such quantities that it was exported. It was widely used as a sweetener, medicine, and to make a drink such as hydromel or mead. The recipe involved diluting one-part honey with four-parts water that was slightly heated. The honey and warmed water were mixed in a "wooden vessel." Then the honey water mixture was put in a pan and brought to a boil until a scum appeared on the surface. This scum would be completely removed. In another pot, hops placed in a linen bag were put into water and boiled until half of the water had evaporated. The resulting

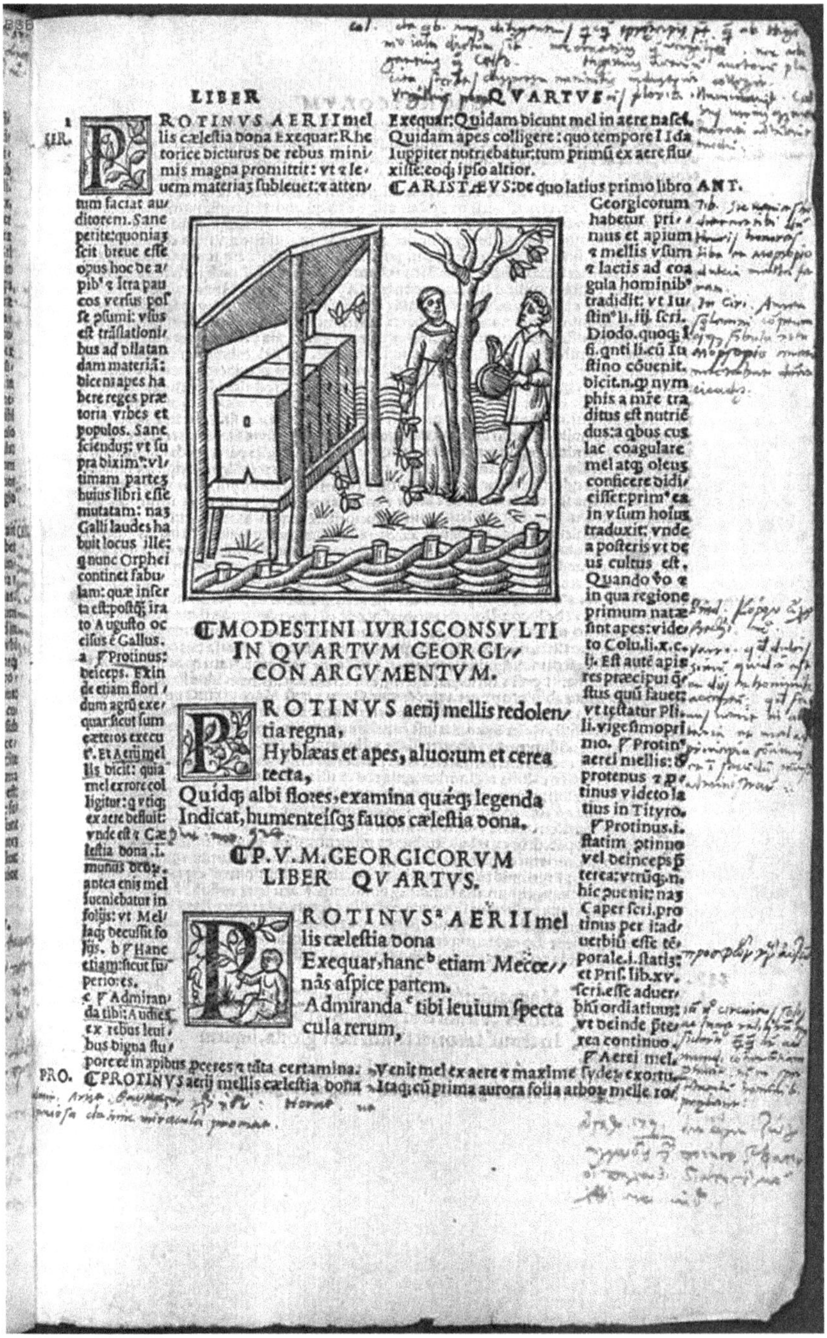

FIGURE 4.6 Early upright box hives from 1507. Virgil. 1507. Georgics, Aeneis. http://daten.digitale-sammlungen.de/0001/bsb00010391/images/index.html?fip=193.174.98.30&seite=313&pdfseitex. Public domain.

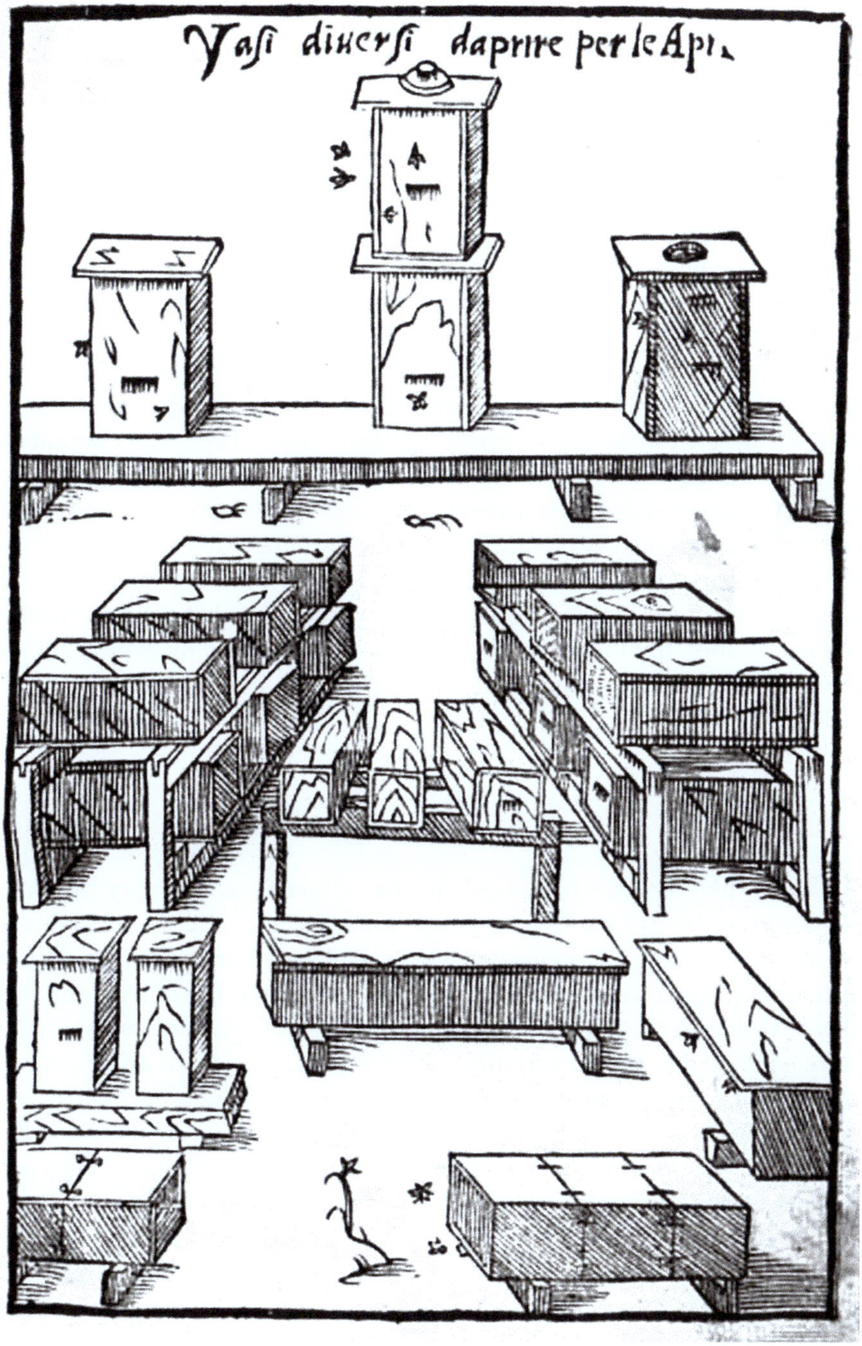

FIGURE 4.7 Italian horizontal and upright tightly fitted box hives. Gallo A. 1596. *Le viniti giornate dell'agricoltura*. E. Imberti, Venice.

bitter tasting liquid and the hops-bag was mixed with the honey water pot and brought to a boil again. Finally, the container was covered with a cloth and slowly cooled until tepid. Next, some dregs of beer (or if beer was unavailable, some baker's yeast) was added. The solution was gently heated until a "pure flavor remains, with a moderate bitterness." Finally, ground ginger and a small pebble were tied into a small bag and suspended by a string in the solution, just below surface. That last step gives the hydromel a more "pleasing taste" (Magnus 1998a).

All of the hives illustrated or mentioned in Magnus are either skeps or log hives. However, box hives were beginning to appear in the literature. A woodcut from 1507 shows four single box hives placed under a lean-to (Figure 4.6). In a 1596 book from Italy is a woodcut showing both horizontal and upright box hives. What is noteworthy in this image is that the upright hives were stacked with an opening between the hives, permitting bees to have a larger space within which they could make comb for brood production and honey storage (Figure 4.7).

Eva Crane wrote that modern beekeeping developed in a series of progressions starting with rectangular box hives, followed by tightly fitted upright box hives, the use of bars, the use of frames, and ending with the careful spacing of the frames in the hives. These two illustrations document that the first two stages in the evolution of rational beekeeping, beekeeping based on experience rather than folklore, had developed by the end of the Renaissance (Crane 1999; Kritsky 2018).

BEESWAX

Beeswax, like honey, continued to be a major commodity during the Renaissance. Its use for sealing letters and candle production continued on from medieval times. Demand for wax was high in Europe. Magnus reported that "many pounds of wax" were used by the Church for candles and for the sale of votives as well as wax figurines of body parts, humans, and animals that were left at a saint's tomb with the hope that it would help to effect a cure. For example, if someone suffered from a bad leg, a small wax effigy of a leg would be used. In 1942, an air raid during the Second World War damaged Exeter Cathedral. While cleaning and repairing the damage a hoard of wax votives was discovered over the tomb of Bishop Edmund Lucy, who died in 1455. The votives were hollow legs and arms, heads, hands, feet, and an eight-inch figure of a woman. There were also animal votives of horses, cattle, and pigs. These figures were clearly cast from a mould, and were affixed to a string, permitting them to be hung on the tomb (Radford 1949; Magnus 1998a).

Beeswax took on a new role as medium to produce small models of larger sculptures. It was common practice for artists to show patrons sketches of intended sculptures. During the early fifteenth century artists started to make three-dimensional models called *bozzetti*. These wax *bozetti* were constructed by shaping the beeswax on a metal armature. One such model sculpted by Annibale Fontana survives and is in the Cleveland Museum of Art (Fliegel 2016).

During the Medieval Age, beeswax was used as a resist to etch the armor of knights. This continued in the Renaissance but led to another application. The armorer Daniel Hopfer around 1490 recognized that his etched designs resembled the engraved copper plates that were used to print images. He applied the same beeswax and cutting technique to flat iron plates, initiating the process of acid washing plates for printing (Houston 2016).

Finally, bee products continued to be used as cosmetics during the Renaissance. Honey was used with black sulfur and alum and applied to hair, which lightened the hair color when exposed to sunlight. Queen Elizabeth I reportedly used beeswax laced with lead to make her vivid red lips more striking from her pale white makeup (Kassinger 2003; Yeola et al. 2018).

SILKS—FIBERS AND TEXTILES

The importance of silk to human culture cannot be overstated; it played a crucial role in many aspects of Renaissance society, including economics, developing technology, art and fashion, women in the work force, and politics. Indeed, the silk industry was an essential driver of the Italian Renaissance, with ripples that extended throughout Europe.

The term "silk" is a general term that describes a combination of two proteins, fibroin and sericin, that together form strong fibers. Fibroin is a polymer of β-pleated sheet protein subunits containing the amino acids glycine, serine, and alanine (~86 per cent), with other amino acids in smaller proportions (Fraser and MacRae 1973: 293–343). The fibroin center, two fibroin filaments, of the fiber is surrounded by sericin, a sticky protein coating. The specific properties of a particular variety of silk depend on the thickness and molecular structure of the fibroin as well as the sericin coating. While spiders and a number of insects produce silk, the use of silk to make fabrics has relied primarily on silk fibers obtained from cocoons of silk moths. Burnham (1980) describes the variety of threads and yarns that can be spun from silk filaments, both long reelable filaments and shorter silk fibers. This practice originated in antiquity, but was not confined to the Far East.

Several different types of silk were produced in Europe and the Middle East, using the silk fibers from cocoons of various silk moths, mostly in the families Saturniidae and Lasiocampidae. These fibers were short and not reelable, as are the silk strands from cocoons of the silkworm *Bombyx mori*, and so were carded from cocoons (similar to the way in which wool is carded), and spun into yarns or heavy threads. The resulting fabrics were known as tusseh (tussah, tussar, tasar) silks. Forbes (1956: 50) describes the likely production of tusseh silks in ancient Greece from two moth species, *Pachypasa otus* (the Syrian silkworm) and *Saturnia pyri* (the giant peacock moth or Aristotle's silkworm) and from other moths in India and elsewhere:

> In India alone more than six species are cultivated including also the Indian "tussur moth" (*Antheraea mylitta*) feeding on the bher tree. Similar species are known to have been bred in China, e.g. the oak-feeding *Antheraea pernyi* which yields the tussah silk of Mongolia. Moths of this type (Antheraea and Philosamia) occur all over the ancient world. It is highly probably that such inferior raw silk was produced from the cocoons of similar insects in the Mediterranean world as "Assyrian" or "Coan" silk before the real Chinese silk was imported in Roman times or its manufacture introduced in the Byzantine Empire.[1]

Peigler (1993) describes the various wild silk moths and the silk produced, and notes that India is and has been a major producer of wild silks, including muga silk, tropical and temperate tasar, and eri silk. A variety of other wild silks have been known for centuries from China, Japan, Greece, a number of African countries, and Mexico. Silk from a pierid butterfly was even used by the Aztecs to make writing paper.

Silk fibers from these moths were carded from cocoons, generally those from which adults had already emerged, and spun into yarns which were heavier than the thread spun from *Bombyx mori* cocoons. These threads and yarns were woven into fabrics ranging from silks used in women's clothing (but having a rougher texture than Chinese silk) to coarser fabrics used in cloaks and other outerwear, as well as velvets and brocades. In addition, this silk was quite suitable to combine with other fibers such as cotton or wool to produce long-wearing fabrics of varying weights and textures. Production of these "wild" silks was widespread across the Indian subcontinent and around the Mediterranean, and it continued in Europe and Asia into the Renaissance (and after).

The extensive use of both tusseh silk, and to a lesser degree Chinese silk, was documented by Abu'l Fazl (1551–1602), a court historian, in his accounting of the court of the sixteenth-century Mughul (Mogul) Emperor Akbar the Great.[2] Among other things, Abu'l Fazl recorded items of clothing used at court with details of their design and composition; twelve particular articles of clothing worn by the Emperor were made of several fabrics and all included silk. Abu'l Fazl also lists fabrics, both gilded (gold-coated thread) and plain, with prices and in most cases, place of origin. Twenty-seven types of silk "gold stuffs" are listed, of which two are identified as being from Europe and one from China. Under a heading of "plain" silks, thirty-nine types of silk fabrics are listed with prices; of those, four are identified as imported from Europe but none are of Chinese origin. Included in these lists are velvets and brocaded velvet, satin, brocades and brocaded silk, taffeta, filoselle silk, silks of a variety of weaves, and some which incorporated wool as well. Most of these came from cities within the Mughul Empire, suggesting that tusseh silk fibers predominated in the silk industry of the time. Blochmann in his 1873 translation of *The Ain I Akbari* from the Persian noted that many of the fabrics listed could no longer be found, having been replaced by English and European fabrics.

However, the fact that a single silk fiber could be "unreeled" easily from mulberry silkworm cocoons, coupled with the fineness and length of the resulting thread, made this luxuriant form of silk dominant in the growing silk industry. The establishment of silk production using *Bombyx mori* in the Byzantine Empire and thence into Europe is a topic for another volume in this series; suffice it to say that this cottage industry was thriving in Europe, and particularly in Italy, by the fourteenth century. Clothing of fine silk was reserved for the aristocracy where it was in high demand, and silk dyed with certain colors, e.g., "royal" purple and later, red, could only be worn by monarchs and princes of the Church. It has been generally agreed that the silk industry in the Venetian Republic as well as in Florence and other Italian cities grew, spread throughout Europe, and served as a critical economic driver of the Renaissance (Molà 2003; Tognetti 2005).

SILK IN RENAISSANCE EUROPE AND BEYOND

Leonardo Fioravanti in 1572 described silk production in Italy in great detail:

> The silk craft is a very noble art, worthy of being plied by any true gentleman, for gentlemen are the ones who use silk; and this craft has very great privileges in all Italy, and in many Italian cities it has its own court of law that administers justice without anyone who interferes, and it enjoys many exemptions and very great privileges, and deservedly so because it is a craft that exalts the rich and helps the poor; and great skills are needed to ply it, since it involves an infinitude of operations; no one is to be found who is capable of doing on his own the many tasks that it involves. The first of which is to hatch the worms that make the silk, and nourish them .... Afterward there are masters and mistresses who, when the cocoons are ready, dry them under the sun for one or two days; and they put a cauldron over a fire and bring it to the boiling point, putting the cocoons inside; and with certain spindles they reel the silk, which when completed goes into the hands of the mistresses, who wind it on the reels; and then it goes to the Spinner, who spins it, and once it is spun it comes back into the hands of the mistresses, who wind it again on the reels

> and bring it back to the Spinner for throwing; and when the silk is thrown, and after the Merchant has checked it, it goes to the Dyer; and the Dyer first boils it in water and soap and then dyes it in any color one wishes; and it returns to the Merchant, who puts it on wooden pegs, with which he spreads it out very carefully and makes it lustrous and beautiful; then it goes to the mistresses, who wind it on certain spools, with which the Weaver prepares the warp he desires and weaves it as he pleases, into velvets, satins, damasks, tabbies, ormesini, and so on; and thus the craft is finished to the glory of God, and to the benefit of the world.
>
> (qtd. in Molà 2003: xiii)

Two centuries earlier, silk was produced primarily by master artisans working in homes, with family members and apprentices assisting. Silk goods were generally made on a piecemeal basis as ordered by members of the aristocracy or clergy. Silk manufacturing for export was confined to the city of Lucca, and the details of manufacture had been held as closely guarded secrets since the twelfth century. This changed when political conflict and violence led to the emigration of skilled artisans from that city to Venice, Florence and other major cities in Italy in the first half of the fourteenth century, breaking the monopoly on the techniques of silk manufacture that had been held by the city of Lucca. These artisans brought with them expertise in production of fine silks in volumes sufficient for trade, and silk industries subsequently developed in the major Italian cities (Tognetti 2005).

Another set of events, however, had an overwhelming impact on Italy and all of Europe: the Black Death in the mid-fourteenth century, with serious consequences:

> ... recurrent plague epidemics, starting with the Black Death in 1348 and the consequent sharp reduction in the urban and rural population, famine, war, the abandonment of villages, the reduction or stagnation of agricultural yields, the growth in wages and hence in the cost of labour, the inability of urban economies to respond positively to changes in demand and consumption, and so on.
>
> (Tognetti 2005: 56)

One of the consequences of this instability was the expansion of the larger Italian cities into surrounding territories, frequently waging war to subjugate smaller cities and thereby becoming much larger states. For example, Florence took over most of Tuscany during the latter half of the fourteenth century, and immediately instituted economic policies that maintained its economic superiority over smaller Tuscan cities. Similar expansion and economic policies were used by other major cities including Venice, Milan, Bologna, and Genoa.

In all of these urban centers, international trade and banking were already well-established, and a new class of merchant-entrepreneurs (setaioli) arose, taking control of silk production by virtue of their organizational abilities, political skill, and capital. They essentially displaced the master artisan-cottage industry system, restricting the artisans to producing piecemeal work and limiting independent production of fabrics for sale. Under the influence of the setaioli, the manufacture and export of silk fabrics increased in volume and was very profitable despite the high cost of skilled labor. Silk manufacturers and silk merchant-entrepreneurs were represented as one of several groups by a professional guild in Florence, the *Arte di Por Santa Maria*, which protected artisans and tradesmen, and it had political clout. Similarly, silk production in Venice was controlled and regulated by the setaioli guild, the *Arte della Seti*, with three guilds of artisans (spinners, dyers, and weavers) being subject to setaioli regulation (Molá 2003; Tognetti 2005). By the early fifteenth century, the importance of the silk industry became evident in political and regulatory policies, such as statutes concerning trade and production in Florence that favored the setaioli within the guild and also goldsmiths who made the gilded silk thread used in expensive brocades and other silks.

Policies of the Florentine Republic itself actively promoted immigration of silk artisans from other cities, offered cancellation of taxes and/or debts to keep artisans in Florence, prohibited silk manufacture in other cities under Florentine control, and even established the death penalty for any artisan leaving Florence to work in one of these cities (Tognetti 2005). As a consequence, the value of silk production nearly doubled during the fifteenth century, as did the number of gold-beating workshops. Quality of silk fabrics produced increased as well and included plush velvets, brocades, and silver and gold brocaded fabrics. The importance of the silk industry in Florence is suggested in a letter written by Niccolò Machiavelli to a friend in 1513:

> As fortune has seen to it that I should know nothing about the art of silk or wool manufacturing, nor of profits or losses, but that I am able to reason about the state, either I must shut up or devote myself to reasoning about this.
>
> (qtd. in Tognetti 2005: 64)

Prominent noble families including the Medici, the Spinelli, the Serristori, and numerous other families also contributed significantly to the growth of the silk industry in Florence by investing in silk manufacturing and gold-beating companies, promoting international trade of high-quality silks, brocades and velvets, and influencing economic policy. Various records, both from companies and from the city, show dramatic growth in the silk industry, as the numbers of workshops of the setaioli increased by half during the mid-1400s (Tognetti 2005). The economic prosperity that resulted from the expanding silk industry through the fifteenth and early sixteenth centuries was no doubt instrumental in the "rebirth" that took place in Florence.

Similar developments occurred in Milan, Venice, and other Italian cities, including protection of the local silk industries and artisans. Molà (2003) traces the development of silk industries, both thread and fabric, in various cities and regions, and notes increases in the numbers of looms as well as the numbers of workers, often in the tens of thousands, as recorded in various documents and census records. In addition to these workers, large numbers of unskilled workers, mostly women, also found employment in various aspects of silk production, from raising silkworms to spinning, sorting silk, throwing and doubling silk, winding thread, and other tasks, an advantage which gave added value to a strong silk industry in any given city. Women were especially important in the spinning of gold thread; Zanoboni (2009: 35) notes that a single shop in Milan employed at least 198 women to wind fine strips of gold, silver, or brass around silk thread. Women were also occasionally apprenticed to artisans who worked in embroidery or other aspects of the silk industry, and they could inherit workshops and invest in silk production as well. Generally, women were excluded from the powerful Italian guilds throughout the fourteenth century, although by the mid-sixteenth century there were more women in skilled crafts, including the silk industry (Brown 1986).

Great importance was placed on the quality of the silk fabrics, especially velvets and other luxury fabrics produced for export by various Italian cities. Regulations regarding the production of various silk fabrics (specific types and weights of silk thread that could be used, thread count, etc.) were instituted, and quality control became institutionalized as part of government bureaucracy. Blends of silk and cotton had been manufactured in the fourteenth century in Florence but had been prohibited in the fifteenth century as inferior, only to re-emerge, along with other blends, in the sixteenth century. Similar evolutions of silk in fabrics occurred in other Italian cities, such as the production in Naples of heavier fabrics for coats, having a warp of silk and a weft of flax or wool.[3] As silk production became established in other parts of Europe, the demand for such blends, including the incorporation of cheaper fibers such as waste silk, wool, and cotton, became more widespread. The Italian silk industry responded by loosening regulations and changing standards for exported fabrics (Molà 2003). In contrast to the use of less-costly silk blend fabrics, courtiers and the aristocracy throughout Italy used fine silk fabrics and velvets embroidered with gold and silver threads and other extravagances to demonstrate wealth and power.

Technology was also stimulated by the burgeoning silk industry, and a variety of technologies emerged in the sixteenth century related to silk production. Hand-driven and hydraulic silk spinning machines made possible the production of large volumes of very consistent threads and yarns, which could produce fabric for local use or exported. Various iron tools associated with spinning, spinning machines, and processes were granted charters or privileges in towns and cities across Italy. Venice began issuing patents for inventions in the fifteenth century, the first general law on patents being passed in 1474. A total of 577 patents were issued in Venice in the sixteenth century, of which about forty were related to the silk industry and ranged from silkworm raising techniques to winding, reeling, and spinning machines to a silk cloth printing machine, as well as patents for specific fabrics. Other patents were proposed for innovations to production that would reduce costs, including the use of clock-style counterbalances to power the silk mills without human labor and another to allow a single worker to power more than one silk mill, eliminating the need for the unskilled workers who did that job. Interestingly, some proposals for patents were denied, being opposed by guilds of artisans and/or judged to have negative economic effects on workers, especially on the poor (Molá 2003).

Sericulture itself had been introduced into Italy in the fifteenth century, with mulberry trees being cultivated in Tuscany. Figuier (1872: 217) wrote that "In 1440 each Tuscan peasant was forced to plant at least five mulberry trees on the land he cultivated." This requirement had been in place elsewhere since the fourteenth century, and similar requirements were passed by various governments, as well as by Pope Sixtus V, who adopted it for all farmers in the Papal State in 1586. Sericulture came to the attention of aristocrats, probably as the silk industry expanded:

> Between 1465 and 1478, for instance, first Lodovico and then Federico Gonzaga, Marquises of Mantua, took great pains to purchase hundreds of young mulberries of the Valdinievole, in Tuscany, for their estate at Saviola, also enlisting specialists of Bologna to plant the trees and to follow their growth.
>
> (Molá 2003: 217)

Plantations of mulberries were established by entrepreneurs and further expanded in Italy during the sixteenth century, despite attempts by various governments to prohibit the export of mulberries to other regions. A consequence of such regulation was theft of mulberry trees; the Venetian Senate responded with a law that such a thief be flogged on market day and then branded, and similar laws were enforced elsewhere (Molá 2003).

Tending the silkworms was largely done by women and children, adding to the rural economy. In 1590, Tuscan raw silk made up about 12 percent of the raw silk used in Florence, and by 1650 about 75 percent came from Tuscany (Brown 1986: 221). Verona and Vicenza were leading producers of raw silk in the Venetian Republic, and sericulture was widely acknowledged as a source of income for both the people and the cities, which gained from a tax on reeled silk. Other states in Italy enjoyed similar success in producing raw silk (Molá 2003).

The success of sericulture in Italy led to efforts to devise ways of producing two crops of silk cocoons per year. In 1567 and 1568, Giovan Battista Guidoboni of Lucca succeeded in obtaining rights to a particular silkworm raising technique; he had also obtained patents for other techniques and devices related to the silk industry in Bologna, Venice, and elsewhere (Molá 2003: 204–6). In 1588, a book titled *Dialogues on the Useful Inventions for Silk* by Maggino Gabrielli (himself a business partner of Guidoboni and the holder of a patent for sericulture from the Pope) described silk production techniques, with the intent of selling his book along with various inventions to prospective buyers at fairs in Italian cities:

> ... his agents "will publicly sell both these books of mine, so that everyone might use them in the silk craft, and all sorts of contrivances and instruments that are needed to carefully tend the little worms, and they will be given at such a low price and for such a good bargain that everyone will profit from buying them, especially because these tools will not wear out for several years". Moreover, at the fairs his agents also had to read aloud the patents granted by the various princes and governments, to inform the populace that no one was permitted to produce the new inventions on his own or buy them from anyone else.
>
> (qtd. in Molá 2003: 209)

Restrictions on emigration of master weavers and other artisans, very rigid in the fourteenth and early fifteenth centuries, became more relaxed, with the result that Italian artisans took their skills to France, Spain, Belgium, Germany, England, and other parts of Europe, although some had done so as early as the thirteenth century. In a number of these countries, silk production was already established; gold spinners and silk workers' guilds existed in Paris, Cologne, Zurich, and London in the thirteenth and fourteenth centuries, many of the artisans having come from Italy (Zanoboni 2009; Farmer 2014). The importance of silk is suggested by Boccaccio's inclusion of Pamphila (who invented silk spinning and weaving according to Greek mythology) in his 1374 *De claris mulieribus,* and an anonymous French translation of that work, *c.* 1440 (British Library Royal Collection), includes an illustration of Pamphila titled "Pamphila collecting cocoons and spinning silk" by the Talbot Master (Figure 4.8).

Montpellier had an active silk trade in the thirteenth and fourteenth centuries (Reyerson 1982) and production of silk fabrics as well as "narrow ware" (ribbon, borders, clerical stoles) began in Paris in the early thirteenth century.

FIGURE 4.8 Pamphila, daughter of Plates, collecting the cocoons of silkworms from trees, and spinning the silk. *De claris mulieribus*, early fifteenth century. British Library, Royal 16 G. V, f. 54v.

By the 1330s, silks from Paris were listed among the luxury silks purchased for the wardrobe of King Edward II of England (Farmer 2014). Tax records from the late thirteenth century indicate that velvets and gold cloth were also being produced in Paris; the silk industry was dependent on obtaining raw silk from Italian merchant-bankers, much of which came from the Mediterranean and Black Sea regions. The Parisian mercers, who were analogous to the setaioli in Italy, owned the raw silk fiber and the finished silk products, and subcontracted the spinning, weaving, and dyeing to artisans (Farmer 2014, 2016). Tax assessment records indicate that the artisan guilds were primarily male, although Farmer (2014) describes several guilds of women in the late thirteenth century who threw silk (throwsters) and wove narrow ware, and Reyerson (2013) notes that in the early 1300s there were five female guilds in the silk industry in Paris.

As in Italy, high standards were applied to Parisian silk production, and regulation of the various steps and processes was largely under control of the mercers.

> By 1324, however, the mercers, with the backing of the provost of Paris, gained broad disciplinary powers over all stages of production of silk textiles and other mercery goods. From then on, the four annually elected *prudhommes* of the mercers' guild had the authority to visit any workshop or shop where silk materials, silk textiles, or other mercery goods were produced or sold. If they discovered any goods or processes (throwing, dyeing, weaving, embroidering) that failed to meet their high standards of excellence, the *prudhommes* had the authority to destroy the goods and charge a fine.
>
> (Farmer 2014: 398–9)

Farmer (2014, 2016) also cites evidence from tax records and court documents showing that some of the mercers of Paris in the late thirteenth and early fourteenth centuries had come from Lucca, Venice, Cyprus, and other areas of the Mediterranean, as well as from other silk-producing cities in France, such as Cologne. However, Paris was not to become the sole center of France's silk industry during the Renaissance; the city of Tours also became home to silk artisans from Italy in the mid-fifteenth century, at King Louis XI's invitation. He also established the city of Lyon as a national center for the silk trade, introducing the manufacturing of silk fabrics in 1466, hoping to replace fine imported Italian silks with high quality French silk fabrics. In 1536 a school for silk winding and spinning was started by Italian artisans.

French sericulture was reported by Figuier (1872) to have begun in Avignon in 1340 when mulberry trees were planted there. De Quatrefages (1873: 667) wrote that silkworms came to France in the "twelfth or thirteenth century" and noted that King Louis XI planted mulberry trees around his Chateau of Plessis-de-Tours and encouraged the raising of silkworms in the mid-fifteenth century. However, sericulture did not gain much dominance until the 1550s when trees were planted in Toulouse, Tours, and several other regions under King Henry II. Sericulture expanded considerably under King Henry IV, who wanted to keep money spent on silk and silk products in France. In the late sixteenth century, the writings of an agriculturist, Olivier de Serres, on the culture of silkworms and mulberry trees came to the attention of the king. The result was the planting of twenty thousand mulberry trees, including in the gardens of the Tuileries, where silkworms were raised. Further, a nurseryman of Nimes, Francois Traucat, promoted the planting of over four million mulberry trees throughout Provence, Languedoc, and neighboring provinces (de Quatrefages 1873: 667). These regions were then able to provide raw silk for the industries of Paris, Lyon, Tours, and other French cities. However, sericulture declined in the years following the death of Henry IV and also later in the seventeenth century following the revocation of the Edict of Nantes (Figuier 1872; Berenbaum 1995).

In Britain, as elsewhere in Europe, silk fabrics were available during medieval times, and were costly, such that only royalty, aristocrats, and princes of the Church possessed silk items. As trade in silk increased in the fourteenth century, statutes were passed in Parliament favoring "all clothworkers of Strange Lands of Whatsoever Country they be," giving them protection and rights to settle where they wished (Warner 1921: 19). Silk is first specifically mentioned, along with wool and linen, in an Act of Parliament of 1363 requiring "artificers, merchants and retail tradesmen" to work in only one class of goods, but specifying that women working as brewers, bakers, weavers, and spinsters working in wool, linen, silk or embroidery were excepted (Warner 1921: 20). Dale (1933: 324) suggests that women in London were the major workers in the silk industry there, based on the fact that by 1368 they as a group petitioned the mayor against a Lombard who had control of "all raw and coloured silks." In the fifteenth century, they sent petitions to Parliament as "Sylkewymmen and Throwestres of the Craftes and occupation of Silkewerk" in an effort to protect their work from foreign competition, the first petition submitted in 1455. Dale (1933) also describes the similarity of the group to other craft guilds, including apprenticeship of young women to established silkwomen, as well as cases in which silkwomen engaged in trade. A number of detailed London records show that silkwomen threw silk (from raw silk by Italian reelers), spun and wove it, producing ribbons, cords, fringe, and corses (to be made into other items and/or embroidered). However, they did not seem to weave large piece goods; such piece goods were evidently imported or produced by Italian weavers in Britain, and the use of silk fabrics also differed in the two countries (Monnas 2009; 2018: 149–85).

The weaving of "broad silk" became established in the late fifteenth century, at least to a limited degree. Warner (1921) describes an Italian weaver, Damico, who was commanded by King Edward IV to weave gold and silver cloth, damask, velvet, and fine silk, and to teach others how to do so; he was given a house in Westminster. Nonetheless, this appears to be the exception, not the rule, as Sir Francis Bacon (here qtd. in Warner 1921: 22), in his *History of Henry VII*, says that all these small articles "the people of England could then well skill to make," but that all other silken fabrics were permitted unrestricted importation, "for that the realm had of them no manufacture in use at that time."

Expansion of the English silk industry, especially broad silk weaving, occurred in the sixteenth century as not only Italian craftsmen immigrated to England but a new group of immigrants from the Netherlands (including Belgium, Holland, and Friesland) settled there. Silk weaving, and most notably tapestry weaving, were major industries in the Netherlands and known for their high quality, and during the armed conflicts of the early sixteenth century, many of these artisans fled to England. In 1561 Queen Elizabeth sent letters to the mayor of Sandwich, giving rights to Flemish immigrant silkworkers there, and otherwise supported the silk industry (Warner 1921). Again in 1585, large numbers of craftsmen and merchants fled to England following the sack of Antwerp by the Spanish, strengthening the silk industry. In addition, the Reformation had a profound effect. The late sixteenth century saw persecution of the Huguenots in France, many of whom were craftsmen in the silk industry. A number of these artisans fled to England prior to the Edict of Nantes in 1599. Huguenot refugees also spread their craft to Germany, Switzerland, and elsewhere. Guilds included both men and women, although there was a late sixteenth-century shift toward the exclusion of women; in Augsberg, Germany for example, female master weavers were only 15 percent of all master weavers by 1600 (Weisner 1986).

Silks had been produced in Spain, and especially in the Iberian Peninsula, since the Middle Ages, and major Spanish cities were centers of silk production and trade during the fifteenth and sixteenth centuries. Valencia was the leading producer of silk following its introduction in the fourteenth century by the Muslims, and one part of the city was known as the Velluters (silkworkers) District; a velluters guild was formed in 1479. Valencia also became a center of sericulture in the sixteenth century, and provided silk fiber to other textile producing cities in Spain, especially Toledo, which then dominated the industry (Lafita et al. 2017). The Spanish silk industry was influenced by Italian artisans who immigrated and brought their skills and knowledge with them, and also by Muslim artisans who influenced styles and patterns of silk (Gasch-Tomás 2018: 251–63). The Spanish would also take sericulture and silk production to the New World.

Silk fabrics were widely traded within the Mediterranean basin, and Ottoman silks made their way to eastern Europe, Russia, and the Middle East as the Ottoman empire expanded. Constantinople was a major trade center for reeled silk and silk fabrics in the Ottoman empire, and major Italian cities had trade outposts in several major Ottoman cities. Ottoman weavers produced a variety of silk fabrics as well. Taxes on reeled silk "contributed significantly to the Ottoman treasury" (Phillips 2018: 107). Damascus and Cairo were centers of silk weaving as well.

In the New World, the Spanish conquest and establishment of New Spain resulted in the spread of sericulture and silk production to the Americas in the early sixteenth century. After several failed attempts to establish sericulture in Hispaniola, mulberry trees and silkworms were successfully raised in the region of Mixteca, and "by 1545 sericulture had reached a commercial scale across Mixteca" (Gasch-Tomás 2018: 255). Mixteca, Mexico City, and Puebla de los Angeles were major centers of sericulture and silk weaving in New Spain, which flourished until the late sixteenth century. By the late 1500s, Ming Chinese silk largely replaced the local silk used by New Spain's weavers, a consequence of Spain's establishment of the Manila Galleon trade route. Sericulture in New Spain declined as a result, but the silk industry in Mexico City and Puebla de los Angeles, as well as trade in fabrics and clothing, persisted well into the seventeenth century (Gasch-Tomás 2018).

TAPESTRIES

In addition to being woven into fine fabrics and velvets, silk also played a major role in the production of tapestries, which were important furnishings of castles and manor houses all over Europe and Scandinavia. Tapestries were usually woven of wool, the finer ones combined with silk and sometimes gold or silver thread. They served multiple purposes, insulating cold stone walls and also providing artistic decoration and communicating power, authority, and magnificence of the owner. While plain wool tapestries were utilitarian, the finer tapestries were works of art, illustrations by figurative composition; they depicted local events or personal scenes, and they were also used to commemorate historical events such as battles, or depict religious scenes. Only the very wealthy could enjoy such tapestries. The finest quality tapestries incorporating silk as well as gold and/or silver thread were most costly; those with silk cost four times as much as wool tapestries, and those with gold or silver thread cost twenty times as much (Campbell 2002: 6).

Tapestry weaving was centered in France and Belgium, especially Paris and Arras, and Flemish weavers were especially skilled in creating complex textures with the use of weaving patterns, colors, and types of thread used (Warner 1921; Campbell 2002, 2008). The designs were colored, full-sized patterns known as cartoons, often made by artists from the Netherlands, but also by other artists, including Italian painters under the patronage of the Medici and other wealthy patrons in Florence. Leonardo da Vinci's *Last Supper* was "translated" into a tapestry in 1514 for Frances I, and Pope Leo X commissioned Raphael to design a set of ten tapestries, *The Acts of the Apostles*, in 1515, which were displayed in the Sistine Chapel (Campbell 2002). Large tapestries with a high warp count (analogous to a high resolution photograph) required many weavers and many months (or even years) to complete, and resembled paintings.

Such fine tapestries were used to demonstrate power and wealth, given as gifts to gain influence, and used as vehicles for propaganda. Campbell (2002, 2008) describes the extent to which this occurred, citing for example the over 2,700 tapestries from the palaces of King Henry VIII that were cataloged following his death. Kings of France and Spain had extensive collections of tapestries as well, and the Habsburgs had sets of tapestries created to show themselves in successful battles and crusades. Philip the Good, Duke of Burgundy, was a leading patron of tapestry weaving, amassing tens of sets of tapestries, some sets of enormous size such as the *Story of Gideon*, eight tapestries with a height of 5.6 meters and a total length of 98 meters. The demand for high-quality figurative tapestries by patrons in the church and the courts of Europe was met by tapestry workshops in Arras, Tournai, Lille, Brussels, Paris, various Italian cities, and even London and Budapest. Huge sums were spent on figurative tapestries by patrons, primarily in Italy but also in the courts and churches of Europe; for example, the set of ten tapestries *Acts of the Apostles* reputedly cost Pope Leo X more than five times what had been paid to Michelangelo for painting the ceiling of the Sistine Chapel (Campbell 2002: 4).

Tapestries thus were an important medium not only artistically, but also to convey in image from philosophical and religious viewpoints, interpretations of historical events and classical literature, and political realities of the fourteenth, fifteenth, and sixteenth centuries.[4]

DYES, INKS, AND OTHER PRODUCTS

Color is a primary means of communication within and between species, and that power of color has been effectively used in human culture. Color of clothing, especially purples and reds, has denoted royalty and aristocratic status, along with having Christological connotations in European culture (Brown 2015: 166–73). Colors have had magical qualities and ritual symbolism attached to them as well; reds for example have signified fire, magical protection, and power. The use of color has also allowed for unparalleled artistic expression, as well as for effective written communication. Both plants and insects have been known for centuries as sources for pigments and dyes used in the making of all sorts of fabrics, cosmetics, as well as paints and inks for writing, illuminating, printing, and painting and sculpture. Their formulations and use expanded during the Renaissance.

Scale insects in the Hemipteran superfamily Coccoidea were used to make dye, and some of the dye-producing insects, the lac scales (e.g., *Kerria lacca*), also produce secretions from which shellac, a resinous liquid, is made. Shellac was used as a protective covering for many objects, from furniture to paintings. Writing inks came from oak galls, produced by several species of oak gall wasps in the Hymenopteran family Cynipidae. These products had major roles in the expansion of art and literature during the Renaissance. In addition, scale insects brought from the New World in the sixteenth century had a major impact on the production and use of red pigments, and on global economy and trade.

Scale insects have been used since antiquity as the source of red pigments (Forbes 1956; Phipps 2012, 2015) valued for dyeing fabrics, as well as use as colorants in inks, watercolors, and paints. A number of scale insects were commonly used in the Old World during the medieval and Renaissance periods. The kermes scales (*Kermes vermilio*, *K. ilicis*, and *K. palestinensis*), known in antiquity, were used in the Mediterranean region as a major source of red pigment used in paints and fabric dyes. These scales feed on several species of oak, one of which is known as kermes oak (*Quercus coccifera*) (Forbes 1956; Phipps 2012). A late fifteenth-century manuscript, the *Manuale di tintura* which describes dyeing techniques of the Venetian tradition, refers to these *Kermes* scales as grain (*grana* in other manuscripts) and also mentions the use of grain barbaresca from North Africa (Bensi 2009). "Indeed, the best-known regions for the production of grain were northern Africa, Spain, the Balearic Islands, southern France, and Greece—particularly the areas of Seville, Valencia, Majorca, Provence, Crete, Corinth, and Patras" (Molá 2003: 110). The Polish cochineal scale (*Porphyrophora polonica*) was used in eastern Europe as a source of red pigment, and it was one of the primary exports of Lithuania and Poland during the fifteenth and sixteenth centuries (Naruta-Moya 2015). The Armenian cochineal (*Porphyrophora hamelii*) produced a red pigment that was used to dye fabrics, as well as by Armenian monks to illuminate manuscripts. These scales were referred to as kermes, and large quantities were obtained by Italian merchants from German and Polish sources or from the Levant (Molá 2003). Dyes from the *Porphyrophora* scales were referred to as *crimsyn* or *cremex* in the *Manuale di tintura* and other sources (Bensi 2009), and were the dyes of choice because of their intense lustrous color. Venice and other Italian guilds and governments wrote and rewrote regulations for dyeing processes in the fifteenth and sixteenth centuries, as kermes was very costly to use, dyes were sometimes substituted and/or misrepresented by unscrupulous merchants, and the competitive market constantly changed.

A different scale insect, the American cochineal (*Dactylopius coccus*), was an important source of red pigment for Indigenous peoples in the New World, especially from Mexico through South America (Phipps 2012; Padilla and

Anderson 2015).[5] The scales were raised on prickly pear cactus (nopal), dried, and used to dye cloth and feathers, and in paints used in documents, with major production in Mexico. Bags of dried cochineal insects, or *grana*, were paid as part of the tribute to Aztec emperors as shown in the sixteenth century *Matrícula de tributos* (Tribute List) and later, to the Spanish (Brittenham 2015). In a 1599 report to the viceroy of New Spain, Gonzalo Gómez de Cervantes described "the raising and planting of the nopal, or prickly pear, cactus that is the insects' host and the seeding, or transplanting, of the cochineal eggs onto the cactus with a *coa*, a brush made of fox hair" (Phipps 2012: 14). Cochineal insects had been raised and their dye used throughout South America from the fifth century onward, and both the accounts of these insects' cultivation and modern genetic analysis strongly suggest that strains were selected for and were in fact domesticated (i.e., could not survive without human care) (Van Dam et al. 2015: 92). Cochineal dyes were present in the fabrics and headdresses of many cultures, including the Chimú and the Inca, as well as in face and other paints. The symbolic power of red was important to the Aztec, Inca, and other Indigenous peoples, both for royalty and in communicating with the gods, just as it was for European royalty and clergy (Phipps [2010] 2012, 2015; Brittenham 2015; Brown 2015).

Following the Spanish conquest and the establishment of New Spain, *grana* were actively collected and sent back to Spain along with other treasures taken from the New World, especially after the discovery that *grana* could be used for dyeing silk:

> In 1537 a silk raiser from Murcia who had emigrated to Mexico, Hernando Marín Cortés, claimed before Viceroy of New Spain Antonio de Mendoza that he was the first person who had successfully applied cochineal to silk thread; it was probably this discovery that opened the way to the employment of cochineal in the textile industries of the Old World.
>
> (Molá 2003: 120)

The cultivation of nopal and cochineal insects was described in some detail in Bernardino de Sahagún's *Florentine Codex*, 1575–1577 (Phipps [2010] 2012; Kerpel 2015). Spain was quick to capitalize on this resource. Phipps ([2010] 2012: 27) states that

> by the mid-sixteenth century the Spanish flotillas that traveled annually between the Americas and Spain were bringing literally tons of the dried insects to Europe. According to Jose de Acosta, a Spanish naturalist who recorded the flora and fauna of Peru, the shipments from Lima to Spain in 1587 included 5,677 *arrobas* (about 144,000 pounds, or 72 tons) of cochineal.[6]

This source of red pigment partially replaced kermes, as the dye produced was a superior red at equivalent or lower cost, although the new dye was not immediately accepted by all guilds and dyers (Greenfield 2005; Phipps 2015).[7,8] In addition to its use in dyeing silk, American cochineal was commonly used to dye wool textiles by the end of the sixteenth century (Molá 2003). Amsterdam and Antwerp, along with Venice, Florence, Rouen, and Segovia, became centers for trade in American cochineal. In the New World, cochineal was an essential part of increasing global trade, as Spanish ships traveled to the Philippines and the Far East (Marichal 2015; Phipps 2015), taking cochineal and bringing back Chinese Ming silk.[9]

Red pigments from scale insects were used not only in the dyeing of fabrics and thread but also in paints used for painting fabric, canvas, and sculptures, and illuminating paper and parchment documents (Greenfield 2005; Wheeler 2009; Padilla and Anderson 2015). In Europe, women could give their skin some blush using rouge, which was made by shearing fibers of red-dyed fabric; these "scarlet shearings" could then be used in small cloths used to rub the face (Wheeler 2009: 58). Red pigments were also used as lip color in Elizabethan England and elsewhere (Greenfield 2005). Cochineal lake and carmine pigments (as well as kermes and lac lake pigments) were also made from dyed silk shearings and were used by painters such as Titian and Coello in the sixteenth century and by masters including El Greco, Rembrandt, Van Gogh, and Renoir in subsequent centuries (Bruquetas and Gómez 2015; Kirby 2015).

Ink used in hand-printed documents was made using oak galls, which are growths made by the tree in response to having cynipid wasp eggs laid in the plant tissue. A number of ink recipes are known (Wheeler 2009; Peverly 2014; Clarke 2016). A recipe for making gall nut ink was recorded by Jehan Le Bégue in 1431 (Yale Library 2013):

> To make good ink for writing, particularly for books.-
>
> Take 4 bottles of good wine, white or red, and 1 lb. of galls, slightly bruised, which must be put into the wine, and allowed to stand for 12 days, and be stirred every day with a stick. The twelfth day it must be strained through a strainer of fine linen, and must be poured into a clean jar, and put on the fire to get hot, until it almost boils. Then remove from the fire, and when it has cooled so as only to be tepid, put into it 4 oz. of gum-arabic, which must be very bright and clear, and stir it with a stick, then add ½ lb. Roman vitriol, and stir it continually with the stick, until all things are well incorporated together, and let it cool and keep for use. And note, that ink made with wine is good for writing books upon the sciences, because, when books are written with it, the letters do not fade, and can hardly be scraped out or discharged from parchment or paper. But if they are written with ink

made with water, it is not so, for they can easily be scraped out, and it may happen that the letters written with it will fade.

4 bottles of wine, or water, or half of each.

1 pound of galls of xij. oz. to the pound.

4 oz. of gum Arabic.

6 oz. Roman vitriol.

And if you took equal parts of each, galls, gum, and vitriol, as much of one as of the other, by weight, it would still be good; as for instance, 6 oz. of each, would be sufficient for the said 4 lbs. of wine or water, or wine and water mixed as before.

Iron gall inks, known from the Middle Ages, were used primarily for writing of manuscripts and other documents, but were also used for drawing during and after the Renaissance. Interestingly, the inks used for printing following the invention of the printing press around 1450 were *not* iron gall inks, which are water soluble, but instead oil-based inks, which adhered better to the metal type (Hirthammer 2019).

Both before and after the printing press came into use, books, manuals, treatises, and pamphlets were produced on numerous insect-related subjects. Books on beekeeping are discussed earlier in this chapter. Maggino Gabrielli published his book *Dialogues on the Useful Inventions for Silk* in 1588, which included information on the raising of silkworms. Two anonymous treatises on dyeing came from Venice and Florence in the late fifteenth century, one dealing with dyeing fibers and the other a more general compilation of information (Molá 2003). The first printed book on dyeing of silk and other textiles was published in Venice in 1548 by Giovanventura Rosetti, *Instructions in the Dyer's Art*; it was very popular and was reprinted several times (Molá 2003; Bensi 2009; Wheeler 2009). Recipes for iron gall inks were included in several books on writing, including Sigismondo Fanti's *Theory and Practice of Writing* (Venice, 1514, the first printed writing book), Ugo da Carpi's *Writer's Treasury* (Rome, 1525), and others (Wheeler 2009: 98–9). Other manuals, treatises, and collections of "secret" recipes and techniques became important sources of information from the sixteenth century on.

Throughout the Renaissance, actual insect products—honey, beeswax, raw silk, scale insect pigment, oak galls—became more valuable as materials and products made from these insect substances proliferated and generated their own value. A great variety of silk and silk blend fabrics, tapestries, inks, dyes and paints and the artistic and printed works made from those inks, medicines and remedies, and technological innovations in beekeeping, sericulture, spinning and weaving impacted the lives of people in many ways. Insect products contributed to what was a monumental impact of insects on Renaissance culture and society.

CHAPTER FIVE

Insects in Mythology and Religion

DARRYL CATERINE

For the first fifteen centuries of the western Christian tradition, scholars and theologians showed little interest in the study of insects—or any other natural phenomenon, for that matter—as an intellectual pursuit worthy in its own right. Rather, observations of the visible world, both natural and human, were valuable only to the extent that they confirmed the Bible's claim of God's sovereignty over Creation, or clarified the ways that human beings should relate themselves to the heavenly realm.

During the age that is now referred to as the European Renaissance (1300–1600 CE), the focus of Christian philosophy and art began to shift from the Creator to the Creation.[1] The religious meanings of insects during this era reflect in miniature these broader shifts in the history of western thought. On the one hand, we find entire villages engaged in debates over the supernatural status of insects in the so-called vermin trials of the sixteenth century, in which the ecclesiastical courts of the Catholic Church heard lawsuits against locusts, worms, and beetles, as well as other such vermin as field mice. Here we can see clearly the continuation of older forms of thought, both Christian and pagan, struggling to make sense of insects and their relationship to humankind.

On the other hand, we also find a small number of artists and writers struggling to think about insects in particular and nature in general in truly unprecedented ways, in approaches that synthesized both Christian concepts and Greco-Roman natural philosophy with careful attention to the forms and behavior of insect life. Three figures stand out in this latter context: German painter Albrecht Dürer (1471–1528); the Flemish artist Joris Hoefnagel (1542–1601); and the English occult philosopher John Dee (1527–1608/9). The fact that all three of these men were from northern European countries is significant, for it was in these areas that the ideas of the Protestant Reformation of the 1500s had swept away both Catholic and pagan assumptions about the natural world, catalyzing new approaches to the study of insect life.

Towards the tail end of the Renaissance we even find the emergence of a proto-scientific approach to the insect world, in the compendia of insects compiled by the Italian naturalist Ulisse Aldrovandi (1522–1605), and the English naturalist Thomas Moffet (1553–1604).[2] While the individual entries of these works include commentaries by both classical and medieval authors, the overarching organization of each book represented a radical break from religious tradition. Aldrovandi and Moffet were encyclopedists in the modern sense of the term; their primary goal was to classify human knowledge of insects as it existed at that time. Most importantly, they were fascinated by insects for their own sake, and whatever older folkloric or moral material appeared in their encyclopedic essays was overshadowed by the authors' sense of wonder in the face of the insect world, as if they were seeing it for the first time.[3]

It was thus during the sixteenth century that insects began to figure prominently in the theological imagination of the Renaissance, in both the ecclesiastical courts of the Church, and the humanistic culture of the northern countries. In spite of these intellectual developments, however, the official teachings of the Catholic Church regarding the natural world were slow to change. The encyclopedias of Aldrovandi and Moffet, as we shall see, reflected the encroaching influence on western European thought of explorations of the so-called New World. It would not be until the seventeenth century that the shock of new discoveries jolted Europe out of its traditional biblical orientation and into the world of modern science. In the meantime, the Renaissance era represents a transitional period in the study of insect life, situated ambiguously between a theological and scientific approach to the exploration of insects.

THE VERMIN TRIALS

A great paradox of the Renaissance is that the era's unmistakable drift towards a more secular state of mind was hastened by an overabundance, rather than scarcity, of supernatural theories claiming to make sense of the world. In the political arena, the emergence of the Protestant Reformation in the early 1500s and the ensuing wars between Protestants and Catholics to establish dominion over western Europe would eventually give way to worldly forms of governance that simply bypassed theology altogether in favor of more rational principles of statecraft. In the sphere of what in sixteenth-century parlance was called "natural philosophy"—meditations on the nature of nature—an analogous competition among vying religious explanations of Creation eventually necessitated a massive restructuring of knowledge around the new methodology of empirical investigation.

Arguments over the religious meaning of insects plays a peripheral but illustrative role in this greater Renaissance debate over the nature of the natural world, and nowhere do we find these arguments expressed more clearly, or developed more dramatically, than in the so-called vermin trials of the age. While the practice of trying agricultural pests in religious courts of law dates back to the ninth century, the number of such cases spiked during the fifteenth and sixteenth centuries (Leeson 2013: 831–3). During this time, close to two hundred class-action suits against vermin—primarily insects—were filed in local ecclesiastical courts throughout Europe, particularly in communities located in northern France, Switzerland, and northwest Italy (Dinzelbacher 2002: 408). Significantly, these were cases initiated by the townsfolk themselves, whose understanding of insect and other animal life incorporated pre-Christian assumptions that these creatures could in some way communicate with and understand the lifeways of human beings, and that vermin could be held morally accountable for infractions against human welfare.

On the one hand, it is not surprising to find such notions of insect sentience among the inhabitants of Europe if we keep in mind that, for many centuries, western Christianity was a complex amalgam of biblical and Greco-Roman concepts first introduced by literate missionaries, and also a wide variety of Indo-European-derived beliefs held by discrete local cultures strewn throughout the territory of the former Roman Empire. The belief that creatures other than human beings possessed a kind of rational intelligence is preserved in the folklore and folk beliefs of Europe. To cite but one example, the custom of a family member informing bees when their keeper had died continued in the British Isles into the twentieth century; in the so-called pagan substratum of European society, the boundaries delineating human and other-than-human intelligences were not always clear-cut (Ransome 2004: 218–22). Furthermore, Christianity on the local level often incorporated pagan elements quite unselfconsciously into its myths and rituals; in the fourteenth century, a Dominican inquisitor sent to east-central France was shocked to learn that pilgrims to the local shrine of Saint Guinefort were in fact observing devotions to the spirit of a greyhound (Chidester 2001: 268–9).

On the other hand, it is remarkable that the institutional Church, particularly during the sixteenth century, would have honored such pagan assumptions about animal life to the point of arranging elaborate trials to ascertain their guilt or innocence. Townsfolk seeking relief from vermin would forward their complaints to the bishop of the diocese; the bishop, in turn, would appoint a procurator, or prosecuting attorney, to represent them in a local ecclesiastical court. For their part, the insects or other pests would also be appointed a defense attorney, and the trial would unfold according to the procedures set for any other legal proceeding. Based on the extant records, insects in these trials met one of two fates: they were either pardoned or declared anathema, meaning that they were excommunicated irrevocably, not only from the local diocese but from all other corners of Christendom as well (Leeson 2013: 817).

The very notion that insects and other vermin could be excommunicated at all represents an even more shocking concession on the part of the Church to pagan assumptions, since such an idea presupposes that these creatures were in some way a part of the Christian fold to begin with. In order to make sense of this apparent absurdity, it is helpful to consider the observation by legal scholar Paul Schiff Berman that the vermin trials represented less of a theological proceeding than a social ritual, overseen by the Church, designed to reassert Christianity's authority in explaining why tragedies (in this case, blight) occur in a universe purportedly governed by an omnipotent and benevolent God (Berman 1994: 316–23). Insects were summoned to court if and only if other, more theologically coherent measures prescribed by the Church had already been taken but failed to redress the problem (Berman 1994: 306–7).

The vermin trials were, in other words, a last resort. Ecclesiastical magistrates were forced to concede that the machinations of insects might be, after all, beyond the ability of theologians to understand fully, but not beyond the power of the Church to condemn. The final sentences of excommunication thus record insurmountable limitations in the extant Christian explanations of nature, even as they reflect the Church's attempt to retain control as the ultimate moral arbiter in a society increasingly marked by religious and metaphysical pluralism.

The first line of defense against insect invasion was penance and prayer, for in a physical universe governed by a providential God, catastrophes like blight were understood to be the result of human sin. It was hardly a stretch for biblical thinkers to find support for such a position; as recorded in the Book of Exodus, God had sent swarms of locust to punish the immoral Egyptians during Moses' and the Israelites' captivity under the pharaoh.[4] Throughout the Middle Ages and well into the early modern era, Catholics and later Protestants as well assumed that God could and would use insects and vermin, as well as drought, to chastise His unfaithful servants. And if collective acts of contrition or devotion failed to remediate the crises, then a second line of defense always lay in store—the possibility of exorcising the locusts or other pests as agents of the devil. In this case, Christians interpreted God's punishment in more severe terms, but retained a belief in the moral coherence of natural disasters nevertheless. A vivid portrayal of this latter position appears in *The Last Judgment* by Renaissance artist Hieronymus Bosch, painted in the 1480s: in the left panel of the triptych, rebel angels tumble towards earth out of the celestial realms, taking the form of insects as they fall. Also reflective of this theological mindset is the nomenclature of "lady bugs"—so named after the Virgin Mary, who in her mercy sent the aphid-eating insects to sinful humanity as a partial remedy for their divine punishment in agricultural woes (Kritsky and Cherry 2000: 18–19).

How, then, were Christians to make sense of blight if neither prayer nor exorcisms worked—that is to say, if the extant theological explanations for human suffering had apparently failed? It was precisely in this context that the vermin trials begin to make sense, as ritual performances in which non-Christian understandings of the relationship between the human and the natural world were duly acknowledged as no less coherent than the Church's position, but ultimately merged with a biblical—though not entirely Catholic—position. What the procurators sought in pressing the punishment of "excommunication" was, more properly speaking, the right to *curse* a species of pest—this latter term being understood as a formal rite of severing the creature's connection with God. In other words, excommunication in the context of the vermin trials referred to the symbolic banishment of the pests from the very fabric of Creation, rather than from the community of baptized Catholics per se—although the theological contradictions implicit in the rite remain to this day.

Procurators could indeed find biblical precedents for the cursing of inanimate objects or other-than-human beings, including God's curse against both the serpent in the Garden of Eden, and against the fortifications of Jericho in ancient Canaan; David's curse against the mountains of Gilboa; and Jesus' cursing of the fig tree in Bethany. Cognizant of the fact that the accursed vermin had been previously identified as agents of God's wrath, the plaintiffs took the greatest pains to justify the leveling of a curse against God himself—a possibility that had been explicitly condemned in the *Summa Theologica* of Thomas Aquinas. The justification of such a blasphemous act, as recorded in the reflections of at least one jurist, was that the *failure* to curse vermin was tantamount to condoning the starvation of human beings, thus incurring an equally mortal sin (Berman 1994: 306).

Implicit in the church-appointed plaintiffs' arguments was the assumption that insects and other vermin had violated a law and were thus endowed with moral agency—a premise that was virtually indistinguishable from whatever pagan understandings undergirded the original lawsuit. Here the case against animal defendants was threefold: first, that each species was governed by laws unique to its own kind; second, that members of each species held each other accountable to these laws; and third, that at least in some cases, an animal's guilt was clearly observable to human onlookers. The justification for the first two assumptions received support from the Greco-Roman ascriptions of human virtues and vices to the animal kingdom, as evidenced for example in the fables of Aesop, where ants were extolled as embodying traits of industriousness; bees, those of sociability, and so forth. Again hearkening back to the Indo-European substratum of European culture, such tales had begun to receive renewed interest in the western world at least as early as the twelfth century.[5]

As for the moral conscience of animals, plaintiffs invoked the behavior of misbehaving dogs, who seemed to register shame before their owners. More generally, plaintiffs could also cite the precedent of trying and executing other domesticated animals in secular courts of law, a practice that was well established by the Renaissance, typically in response to taking the life of a human being. While never cited in the vermin trials, the moral culpability of individual animals even had biblical justification, as in the Torah's prescription to stone to death an ox that gores a person.[6]

For their part, the attorneys who were appointed to animals mounted defenses of their clients that in turn reinforced the notion of their sentience, and dismissed the allegations of plaintiffs as absurd. On the one hand, records exist of a lawyer pleading to excuse a colony of field mice for failing to appear in court because a clowder of cats had stood in their way (Leeson 2013: 812). On the other hand, defense attorneys rejected in principle the very premise that insects or other creatures possessed moral agency—citing, once again, Thomas Aquinas as an authority—and thus dismissed the charges against them as absurd. The only law that bound the animal kingdom was that of survival and self-propagation, which for the defense was implicitly mandated by the blessing God had imparted to all living creatures in the Book of Genesis (Berman 1994: 311).

Whether or not insects were excommunicated or received pardon seemed to depend on whether or not they continued to pester the farmers. As one modern-day economist has argued, the Church would have had practical incentives to make sure the trials were just long enough to allow the vermin to leave of their own accord, while still giving the impression that their exodus had something to do with the pronouncement of excommunication (Leeson 2013: 813). From this perspective, the purpose of the vermin trials was to demonstrate (by ruse) the supernatural efficacy of excommunication, which could then be held up as a potential punishment for any congregant who refused to observe the laws of the Church—specifically, for those who failed to tithe (Leeson 2013). There is indeed some merit to such an analysis. After all, if the trials failed in all cases to exterminate the pests, then it is difficult to explain their increasing popularity in the late Renaissance—and barring any supernatural explanations, the probability of the vermin leaving a field on their own accord would certainly increase over time.

At the same time, religious communities do not abandon their faith simply because their prayers go unanswered or their rites fail to produce tangible results. While there is no data to prove one way or another how often the vermin trials "worked," there is enough evidence in the substance of the trials themselves to suggest that their primary purpose was to find meaning in the calamity of agricultural blights, rather than to effect a pragmatic outcome. As we have seen, none of the jurists in the trials attempted to resolve the question of how it is possible to excommunicate a species that is not part of the Christian community to begin with. To excommunicate insects did not mean, in this context, to make them go somewhere else in this world. Rather, to level a curse against insects was to label them as godforsaken, and to render them completely outside of any coherent sense of moral order. In this sense, an accursed creature assumed a status not unlike the present-day notion of a "terrorist," banished to the realm of total otherness, beyond the pale of God and the devil alike. Thus what began as a concession to pagan notions of insects as at least partially intelligible and intelligent beings ended as an indictment of their radically alien nature.

As we will discuss more below, the vermin trials thus encapsulate in miniature a broader trend during the Renaissance in which older frameworks for understanding insects—Christian, pagan, and Greco-Roman alike—fall away altogether. The Church's attempt to narrate the nature of nature in the case of the vermin trials succeeded as an emotionally and aesthetically dramatic display of power, but one that lacked intellectual coherence. In the end, the competing worldviews articulated during the trials only served to relativize each other's assumptions, with the verdict of (cosmic) excommunication failing to synthesize or integrate them. For a more successful attempt in combining Christian, pagan, and Greco-Roman understandings of insects, we now turn to the story of how the scarab beetle, once a central religious symbol of the ancient world, was rediscovered and interpreted anew in the closing centuries of the Renaissance.

THE NORTHERN RENAISSANCE: ALBRECHT DÜRER

If there is one image in Renaissance art that captures a truly novel approach to the study of insect life, it is the untitled 1505 drawing of a male stag beetle by Albrecht Dürer (Figure 5.1). Painted in brown and charcoal-grey gouache, the beetle stands by itself against a light watercolor wash background, with no other images to distract the viewer's gaze upon its lifelike form. Enlarged to fill most of the paper, which measures approximately 5.5 × 4.5 inches, the exquisitely reproduced details of the insect are plain for all to see: the tripartite body; the large, clawed pincers atop a slightly tilted head; the antennae; and the six, tarsus-tipped legs bedecked with coarse hairs.

Renowned even in his own day throughout Europe, Dürer, a German artist and writer, is today considered one of the masters of the Northern Renaissance. This latter term refers to the flowering of Christian humanism[7] in the countries beyond Italy, vast areas of which broke away from the Catholic Church in

FIGURE 5.1 Albrecht Dürer (German, 1471–1528), *Stag Beetle*, 1505, watercolor and gouache; upper left corner of paper added, with tip of left antenna painted in by a later hand. 14.1 × 11.4 cm (5.5 × 4.5 inches), 83.GC.214, The J. Paul Getty Museum, Los Angeles.

the wake of the Protestant Reformation in the early sixteenth century. It is primarily to those northern regions of Europe, as well as England, that we shall now turn, for it was in what are now the countries of the Netherlands, Germany, and England that insects in general, and the beetle in particular, took on new religious meanings not found in the Catholic vermin trials.

As we have discussed thus far, the religious meanings of insects cannot be appreciated apart from the broader theological understandings of nature, and the contributions of the Northern Renaissance to this discussion are no exception. Before returning to a discussion of Dürer and his like-minded contemporaries, it is necessary first to make some preliminary and general remarks about the differences between Roman Catholic and Protestant notions of the natural world and its relationship to God. In the first case, Roman Catholic theology, from the second-century Church Fathers down through today, has been guided in its reflections on nature by what we might call a sacramental understanding of Creation. In this view, God's presence pervades and permeates the people, places, and things of the natural world. While God became fully manifest in the person of Jesus, God's power and glory are also made visible in the various particularities of Creation.

Because of this understanding, Christianity before the Reformation had little difficulty in incorporating the sacred places of pagan societies, or embracing the sacred times of their festivals, into its theological worldview. Most churches and shrines of Catholic Christendom were built atop pre-Christian places of devotion, while most of its holidays, including feast days for the saints, were timed to coincide with previously observed ceremonies. Rather than destroying paganism, Catholicism transformed it, teaching that the natural world was not an end unto itself, but rather a sacramental link to God.[8]

In stark contrast to Catholic sacramentalism, Protestant theology emphasizes to varying degrees the transcendence of God above and beyond His creation. Starting with the teachings of Martin Luther and continuing into Protestant theology today, the incarnation of God in the natural world, in and through Jesus, was a singular and exceptional event. Outside of this revelation, the manifestations of God in the natural world are either rare or non-existent. For this reason, the Reformation proceeded in Europe not simply via the dissemination of ideas, but also through physical destruction of Catholic holy places, and the condemnation of many of its ceremonial observations. The most extreme reformers cast Catholicism not as a pre-existent form of Christianity, but as a religion of "paganopapism"—a pejorative term reducing the Catholic faith to an unholy mix of papal worship and pagan practices (Chidester 2001: 345–6).

Some scholars have suggested that the Protestant Reformation, in its iconoclastic stance towards the Catholic sacramental world, paved the way for a thoroughgoing secular view of nature as lacking any religious value in its own right.[9] As plausible as this argument may be, the fact remains that nature continued to be viewed as a sacred realm even within the Protestant countries of northern Europe for at least one hundred years after Martin Luther first set the reforms in motion. And as it so happens, the fate of insects during the period of the Northern Renaissance clarifies exactly how a small but influential group of artists, philosophers, and theologians reconceptualized nature somewhere in between a sacramental and a desacralized understanding of Creation, which returns us to Dürer's print.

Gazing upon the stag beetle some five hundred years after it was painted, it is tempting to conclude that Dürer was nothing more or less than a modern entomologist. After all, the image stands all by itself, devoid of either commentary or imagery that might render its meaning immediately comprehensible—it seems to be "just an insect." In fact, it was this lack of an explicitly Christian context that made the print so radical in its own time. Medieval and Renaissance artists had been depicting insects for centuries in many forms, including in books of herbal lore, in the margins of illuminated manuscripts, and even in devotional artwork featuring such biblical figures as the Virgin Mary and the Christ child (Pacht 1950: 13–47). In these settings, insects were illustrative of other things: in religious works, they served as emblems for various virtues and vices; or in the case of flies, they signified decay and mortality. A famous example of this latter meaning can be found in Carlo Crivelli's painting of the Madonna and Child, dated to approximately 1480, in which the infant Jesus gazes bemusedly upon a fly resting in the lower left-hand corner of the frame. Here the meaning of the insect could be understood in juxtaposition to the images of the Holy Family, as an image of lowliness and degradation that clashes with the splendor of the divine—and also, perhaps, as an emblem of Jesus' mortality.

In stark contrast to such images, there is no biblical frame of reference in Dürer's painting—and for the first time in the history of western art, an insect is featured as the sole subject, rather than a secondary image in the margins. Considered in relation to the artist's other works and against the backdrop of its own time, Dürer's print records a moment in the Northern Renaissance when Christians began to theologize anew the religious import of the natural world. In the most general terms, this new theological turn rejected the sacramental notion that God was present in nature, but found in nature ample clues of God's transcendent glory; nature was, after all, His handiwork, and could be studied as a reflection of divine intelligence in its own right. If sacramentalism imagined the divine presence descending from heaven to earth, then this new orientation exhorted Christians to redirect their attention upwards from earth to heaven, availing themselves of nature as their guide.

For Dürer, who did not leave the Catholic Church but was nonetheless profoundly influenced by the ideas of the Reformation, the ultimate value of nature lay in its beauty. "Art is omnipresent in nature," he once wrote, "and the true artist is he who can bring it out."[10] The stag beetle of 1505 was in fact only one of his many studies of natural scenes and animal life, an enlargement of the same insect that had appeared in the corners of two earlier paintings. In these previous works—*The Adoration of the Magi* (1504), and *The Virgin Among a Multitude of Animals* (1503)—the beetle did not interact in any way with Christ or the Virgin, as it did in Crivelli's piece. Rather, in both paintings its frame of reference was the natural world itself, portrayed especially in his homage to Mary as a veritable cornucopia of vitality, manifesting in a diverse array of flora and fauna.

By lifting out and accentuating a form of life that was deprecated in its own day, Dürer was perhaps asserting his own conviction that God's aesthetics, as revealed in the natural world, surpassed the ability of humans to understand fully. In the artist's own words, "never imagine that you either can or will make anything better than God has given power to his creatures to do. For your power is weakness compared to God's creating hand."[11] Or, as one art historian has summarized the sentiment, "beauty lies even in humble, perhaps ugly things, and the ideal, which bypasses or improves on nature, may not be truly beautiful in the end" (Kleiner 2015: 629).[12] While falling short of what we might call a fully developed theology of nature, such musings nevertheless exemplified how Reformation thinkers like Dürer were attempting to understand divine beauty anew, relying on close observations of nature, rather than biblical ideals, as their guide to theologizing.

Whether or not Dürer's stag beetle carried more overt religious or even Christological meanings, either to himself or to his audience, is more difficult to ascertain, as the artist himself wrote nothing on the subject. Still, the question remains: why a beetle and not another insect to showcase the insect world? One could speculate that the beetle, particularly the male stag beetle, figured prominently in German folklore as an animal associated with Thor, the protector of humankind and the god of thunder and lightning (Ratcliffe 2006: 92). Less speculatively, we do know that Dürer was intimately familiar with Horapollo's text, the *Hieroglyphica,* composed in the fifth century and rediscovered in 1419. As its name suggests, the *Hieroglyphica* is a compilation of Egyptian hieroglyphs, together with their purported meanings. One of these entries, entitled "How an Only Begotten," discusses at length the glyph of the scarab beetle: "To denote an only begotten, or generation, or father, or the world, or a man, they [the Egyptians] delineate a scarabaeus. And they symbolize by this an *only begotten* [Greek: monogenesis] because the scarabaeus is a creature self-produced" (Horapollo 1840).

When Dürer first came in contact with the *Hieroglyphica* is unclear; one scholar surmises that it could have been as early as 1500, five years before the painting of the stag beetle (Cambefort 2006: 201). In any case, together with the Nuremburg scholar Willibald Pirkheimer (1470–1530), Dürer prepared a series of illustrations for a German translation of the work, presented to the Holy Roman Emperor Maximillian I (1459–1519). His familiarity with the chapter discussing the scarab is assured, as he crafted a depiction for it. Curiously, however, Dürer chose to represent the insect as a millipede, for reasons that remain obscure to this day.

With the Renaissance discovery of the *Hieroglyphica,* western Europe rediscovered the ancient religious significance of the scarab or dung beetle. Horapollo's entry describes in detail how the Egyptians associated the insect with immortality—in part due to their perception that the insect was born through spontaneous generation; and in part because they associated the beetle's rolling of dung into small balls in an east-to-west direction with the sun, a symbol in ancient Egypt of eternal life. The symbolism of the scarab spread from Egypt throughout the rest of the Mediterranean world during the Roman Empire as signifying both divine protection and invincibility in warfare (stemming from associations between armor and the beetle's hard exoskeleton). In the fourth century it was also taken up as a Christian symbol by Ambrose of Milan, due in no small part to the fact that the Greek word for self-begotten (monogenesis) was the same term used in the Gospel of John to describe Christ. A vehement critic of all things pagan, Ambrose nevertheless appropriated the scarab as a Christological symbol par excellence, writing that

> [Jesus Christ] was on the cross like a worm, on the cross like a scarab. But how good a worm attached to the wood, how good a scarab shouting from the cross .... How good a scarab who changed into virtue the shapeless ordure of our body, how good a scarab who exalted the poor from his dunghill.
>
> (qtd. in Cambefort 2006: 203)[13]

It is unclear, once again, whether or not Dürer was familiar with the full extent of the scarab's conflation with Christ in the ancient world, but the associations with Christ as the only-begotten Son would have been rather self-evident. On the one hand, it seems incongruous that with such an understanding Dürer would have chosen to portray a male stag beetle, rather than a scarab, if his intentions had been to encode the ancient Christological meaning into his painting. On the other hand, it is equally plausible that he saw these two species as interchangeable, particularly given the stag beetle's association with divinity in Germanic culture.

THE NORTHERN RENAISSANCE: JORIS HOEFNAGEL

While the precise theological meanings of insects in Dürer's work remain partially developed or ambiguous, they do foreshadow more explicit formulations in subsequent works of art and literary texts of the Northern Renaissance. The direct heir of Dürer's entomological metaphysics was the Flemish miniaturist Joris Hoefnagel. Between the years 1575 and 1582, Hoefnagel crafted for an unknown patron *Four Elements*, a four-volume set of some 277 miniatures, painted in gouache and watercolor, illustrating a panoply of animal life. In 1590 or 1591, the work was acquired by the Holy Roman Emperor Rudolf II, who added it to his massive museum in Prague. Aside from being a great admirer of all things wondrous and occult, Rudolf was also enamored of Dürer's oeuvre, and served as patron to many artists whose works were inspired by the German master's style—including Hoefnagel—thus inspiring what has come to be known among art historians as the Dürer Revival.

The *Four Elements* is not only a stylistic continuation of Dürer's work, however. More precisely, it is an elaboration upon the theological themes, both natural and Christological, already implicit in his portrait of the stag beetle. As its title suggests, *Four Elements* organizes the fauna of Creation according to their associations with the substances of fire, earth, water, and air—the elements which, up until the modern era, were believed to be the fundamental building blocks of nature. Accordingly, the folios of Hoefnagel's work are entitled, in order: "Fire. Rational Animals and Insects"; "Earth. Quadripeds and Reptiles"; "Water. Aquatic Animals and Shellfish"; and "Air. Birds and Amphibians." Of these four volumes, it is the first, featuring Dürer-inspired lifelike recreations of insect life, that most clearly explains the overall theological import of the work. In short, insects are herein presented as the most wondrous of all of God's creatures: they are, in their paradoxical mixture of beauty and strangeness, aids for contemplating the unfathomable creativity of the Creator.

But there is more to it than this. Hoefnegel's pairing of insects with the energizing element of fire builds upon the medieval and Renaissance understanding of matter as a both dynamic and generative milieu that was constantly giving birth to new things. Alchemists, to cite one example, believed that metals gestated and grew to completion in the incubating womb of the earth (Eliade 1978: 40–2). Artists, to cite another, understood their various media as endowed with vitality, and their craft as learning to work in harmony with matter's own intentions (Smith 2006: 114–23).

But the most poignant example of nature's fecundity was arguably the belief that some forms of life, including plants and many insects, arose spontaneously from the earth, water, or putrefying vegetable and animal matters. Aristotle, the towering philosopher of the medieval and early Renaissance world, had committed the theory to writing in his *History of Animals*, the culturally authoritative text on fauna up until the early modern period. It held such sway that the naturalist Francesco Redi, writing in the seventeenth century, was careful to couch his disputations of the spontaneous generation of maggots in theological language, lest his experiments awaken the condemnation of the Church's Aristotelian scholars.

Hoefnagel's classification of insects as creatures associated with fire derived from his implicit acceptance of Aristotelian thought in two ways. The first was that heat was the agent that catalyzed spontaneous generation from either earth or water; insects were thus fire's direct offspring. In Book III, section 11 of his treatise, "On the Generation of Animals," Aristotle had clarified:

> Animals and plants come into being in earth and in liquid because there is water on earth, and air in water, and in all air there is vital heat so that in a sense all things are full of soul. Therefore living things form quickly whenever this air and vital heat are enclosed in anything.

Beyond this, fire was also seen as the element that both generated and destroyed itself; thus for Hoefnagel, self-generated insects were analogues of fire. In the words of one scholar, "insects do not live in but rather embody their element and are thus deeply one with nature" (Hendrix 1995: 382). In fact, with the exception of its title page, the first volume of *Four Elements* contains no images of fire, whereas the other three volumes portray their select animals living on or within the elements of earth, water, and air.

As for the "rational animals" accompanying insects in the fire volume, the "specimen" in question was a sixteenth-century man named Pedro Gonzalez (or Petrus Gonsalvus) who was a hirsute, born with the condition of *hypertrichosis universalis*, which causes the abnormal growth of hair over the entire body. Born in the Canary Islands, Gonzalez was taken from his home to the court of the French King Henry II in 1547, where he was educated in the humanist tradition; he was later moved to the Netherlands, where he married a woman who was not a hirsute. Together they had seven children, four of whom inherited their father's condition. Finally, Gonzalez and his family were relocated to Italy, where Ulisse Aldrovandi met them, and sketched Gonzalez's portrait. This was later featured in Aldrovandi's posthumous publication, "On the History of Monsters."

In keeping with the assumptions of his day, Hoefnagel featured paintings of both Gonzalez (standing alongside his wife) and two of their hirsute children, as his examples of "rational animals"—hybrid creatures thought to be betwixt and between two orders of creation. Considered together, hirsutes and insects were intended to inspire the viewer to contemplate the truly transcendent, even alien, designs and will of nature's Creator. Such a theological point of view was reinforced by an epitaph inscribed above the heads of Gonzalez and his wife. This read, "Of all the miracles made by man a greater miracle is man; among those visible miracles, the greatest is the world; among those invisible, God; if we see that the world exists we believe God exists." If Dürer had intimated that the beauty of nature points beyond itself to the splendor of God, here Hoefnagel stated clearly that the wonders of nature point beyond themselves to the holy otherness of nature's Creator.

The species of insects featured in the fire volume were manifold, including various kinds of butterflies, scorpions, spiders, and flies. But in keeping with his admiration for Dürer, Hoefnagel was also careful to include a portrait of the male stag beetle, in an exquisitely precise reproduction of the original. Here at last we find an explicit reference to the Christian symbolism of the insect, written on the page opposite the image: "The beetle armed with horns generates itself; the man Christ was himself his sole origin"—in a clear allusion to the "self-begotten" nature of both. While the artist gives no indication that the stag beetle was somehow privileged above other insects for its association with the Savior, the inclusion of this sentence does suggest that, by Hoefnagel's time, the Christological significance of the beetle had become public knowledge, or was at least known among a smaller group of humanists, including the Holy Roman Emperor himself. At least one scholar has even suggested that the Christological meaning of the beetle passed over into the still life paintings of seventeenth-century artists in Germany and the Netherlands, among Protestant artists "who no longer recognized saints and did not accept representations of God, Christ, or the Holy Spirit" (Cambefort 2006: 215). Modeled not on *Four Elements* but on a number of Hoefnagel's still lifes featuring flowers and insects, these later works typically included beetles inconspicuously tucked into the margins, in what may have been an implicit acknowledgment of their association with Christ (Dicke 2000: 228–37).

For Hoefnagel himself, however, the theological import of insects seemed to have had less to do with doctrinal orthodoxy than with the larger issue of reconceptualizing the relationship between nature and God. Hoefnagel's own institutional affiliations with Catholicism are unclear. He never broke ties with the Church, and took on commissions to produce Catholic devotional art; at the same time, he was well acquainted with a number of Calvinist reformers in the Netherlands as well as France, and his own father was a Calvinist. What is clear is that his life was profoundly affected by the religious warfare between Catholics and Protestants; when Dutch Protestants rose up against their Spanish Catholic king in the Eighty Years War (1568–1648), Hoefnagel fled into permanent exile from his native country.

Theological disputes over Catholic sacramentalism were not the only factors catalyzing new considerations of nature. Especially in northern Europe and England, where the Wars of Religion raged, the vision of re-establishing Christian theology on the basis of nature became attractive for political reasons as well. Nature was potentially a common ground for theological speculation among Catholics and Protestants alike. It seems very likely that Hoefnagel's attraction to the world of natural wonders was motivated precisely by such concerns, even though his reflection on insects revealed a distinctively Protestant cosmology wherein God dwelt distantly from His creation (Hendrix 1995: 389–90).

THE NORTHERN RENAISSANCE: JOHN DEE

The final context for considering the religious meaning of insects in the Northern Renaissance is the renewed interest in the ancient art of alchemy that flowered during the sixteenth century. Here we are provided with yet another example of how the beetle—here the scarab of ancient Egyptian fame—reflected contemporaneous efforts to re-conceptualize the relationship between the natural and supernatural world more broadly. Reference to the scarab appears in a work by one of the most famous alchemists of the era: John Dee, an English polymath, advisor to Queen Elizabeth I, and an enthusiastic proponent of establishing a new British empire, not on religious tenets but on the principles of natural philosophy.

The alchemical worldview in which Dee enshrined the scarab can be traced back, in the West, to the world of late antiquity. Stated most concisely, alchemy is an art whose goal is to transform nature from its perishable state to an incorruptible one. In practice, alchemists of the pre-modern world combined what today is called chemistry, on the one hand, with religious contemplation or meditation, on the other: their desired goal was at once to transmute base metals into gold, and ordinary human nature into immortal spirit. Consequently, modern scientists have been able to reproduce from ancient alchemical texts bona fide chemical reactions, while modern psychologists have found in the same texts insightful reflections on the spiritual nature of the human person.[14] In the pre-modern context, however, natural philosophers did not distinguish between science and spirituality. This was because they assumed that all parts of the universe are interdependent, meaning that a change in one part would effect a change in another, so that work on transmuting physical substances would then cause an analogous change in human nature, and vice versa.[15]

Dee's alchemical magnum opus was a small tractate of theorems entitled *The Hieroglyphic Monad*, and it is in the eighteenth chapter that the scarab appears, once again, as a prominent religious symbol of the Renaissance theological imagination. Here the Elizabethan alchemist presents his own variation of Aesop's fable of the Eagle and the Beetle, a well-known tale in the Renaissance because of its inclusion in a compilation of Greek and Roman proverbs collected by Erasmus of Rotterdam, the famous Dutch Renaissance philosopher, and published in 1500. In Aesop's version, the story relates that an eagle was one day chasing a hare, who dived for protection into a dung beetle's lair, but was captured and eaten nonetheless, overriding the beetle's own protests.

As the story continues to relate, the scarab found the eagle's nest and destroyed its eggs in retaliation, an act that prompted the eagle to fly up to Zeus's abode on Mount Olympus and beg for protection. The god obliged, placing a new bunch of eggs on his lap, but the beetle was not to be outdone. It too buzzed up to the top of Olympus, where it dangled a ball of dung in front of Zeus, who in shock and disgust quickly stood up from his seat, letting the eagle's eggs smash on the ground. The beetle then related its side of the story to the god, who tried in vain to assuage its anger. Finally conceding, Zeus declared that from that time forward, the eagle would always lay its eggs in the season when the scarab was inactive.

In Erasmus' hands, Aesop's fable took on a decidedly political moral, in keeping with the humanist values of the Renaissance: the eagle stood in for despotic rulers, while the beetle for their powerless but valiant subjects (Cambefort 2006: 209) In Dee's retelling, however, the fable became an occasion for conveying some of the central tenets of alchemy. In the eighteenth chapter of *The Hieroglyphic Monad*, the egg figured as a symbol both for the generative powers of nature, and also for the alchemist's vessel, or cucurbit, in which various chemicals were "gestated" to produce an elixir known as the Philosopher's Stone—a mystical substance that would turn base metals into gold, and confer immortality to the alchemist. The scarab, in turn, was renamed by Dee as the "heliocantharis," or solar beetle, which appears to refer to the alchemist himself, as Dee designated alchemy as the "art of the scarab" (Dee 2000: 26). In this case, the beetle in the *Hieroglyphic Monad* stood in for the mythical phoenix, a bird who, like the archetypal alchemist, rises to new life after dying to his old form. According to legend, in the last hour of its life the phoenix builds a nest and burns it, and from its ashes rises a worm, which then transforms into a new bird. Its first task, post-resurrection, was to bring the remains of its nest, as well as its former body, to Heliopolis, literally the "city of the sun" or the abode of the solar deity.

Interestingly, a similar conflation between the scarab and the phoenix can be seen in reproductions of the *Four Elements* done by Joris Hoefnagel's son, Jacob (1575–1630). Issued in 1592, the reprint of Hoefnagel's iconic scarab (which, as we have seen, is a reproduction of Dürer's original) is now featured with its wings open, and situated next to a scorpion, included as a species of a stinging insect. Alongside this portrait is a slightly modified allusion to the beetle's self-begotten nature: "Neither a male engendered me, nor a female conceived: I am myself the sole author of mine *and my breed* [italics added]." Given the obvious point that Christ did not reproduce, the question arises as to what exactly Jacob Hoefnagel meant by "breed." In light of the phoenix's exceptional way of resurrecting from its own self-generated offspring, the mythical bird is a likely answer. This intriguing possibility suggests that Dee's distinctively alchemical interpretation of the scarab had become part of the symbolic *lingua franca* among Prague's alchemical court culture (Cambefort 2006: 212).

Given Queen Elizabeth I's support of alchemy in Germany (not to mention that of the Holy Roman Emperor, Rudolf II, who patronized Dee's pursuits), the sixteenth century was truly the heyday for alchemical pursuits. With the deaths of such supportive leaders, alchemy soon fell out of official favor, becoming associated in both Catholic and Protestant countries with heretical or even demonic teachings. This development was in many ways a loss for Christian theology, insofar as grand syntheses like those of Dee represented the highpoint of attempts to bridge the gap between God and Creation, and faith and reason. As we have seen, the ever-changing symbolic significance of insects accurately reflects this broader theological pursuit. After the 1500s, as the gap between God and nature grew ever wider, insects accordingly diminished in religious importance as well, until they became simply the modern "pest," devoid of wonder and associations with divinity.

The sole exception to the waning of theological interest in insects was a return to the Christological meaning of the scarab found in the writings of the German Jesuit polymath, Athanasius Kircher (1602–1680). Writing in 1636, Kircher returned to the conflation of the beetle's body with the body of Jesus:

> What does the beetle's body mean, if not the Unique Son whom the Father has established as principle, rest, and end of everything, by whom everything is done, and without whom nothing is done? ... Let those who get nauseous by this comparison of the vilest, basest, most horrible and most stinking of all beings, remember that this merely expresses the human condition's baseness, which God's infinite majesty has accepted to dress.
>
> (qtd. in Cambefort 2006: 216)[16]

While some modern historians have characterized Kircher as a late "Renaissance man" for his wide range of interests, his undisguised disgust of the scarab beetle, and of humanity itself, suggests exactly how much was lost with the passing of sixteenth-century theological reflections on nature and the divine nature of the human being.[17] As attempts to bridge heaven and earth receded into the past, Christianity abandoned an important language for articulating the sanctity of both Creation and its creatures—human, insect, and all other life forms as well. As the Renaissance gave way to early modern and Enlightenment eras, the gap between God and the cosmos first articulated by the Protestant Reformers became increasingly formidable. This is not to suggest that Albrecht Dürer, Joris Hoefnagel, or John Dee offered foolproof metaphysical answers; it is rather to suggest that their questions about the sanctity of nature in general, and insects in particular, were rooted in concerns that still haunt many today.

INTO THE NEW WORLD

Our discussion of insects in the Renaissance has thus far proceeded as a foray into the meaning of nature as it was conceptualized by, for, and about western Europe. The worldviews in which insects took on significance and meaning in the sixteenth century were indeed shaped primarily by sources that were either derived from the western past—the legacies of Christianity, European paganism, and Greco-Roman philosophy—or else crafted anew from innovative interpretations of the same traditions, by artists and thinkers of the Italian and Northern Renaissance. But in addition to these intellectual currents, there was yet another development that profoundly changed the way that Europeans of this age thought about themselves and the natural world. This was, to borrow a phrase from the historian Anthony Grafton, the "shock of discovery" that reverberated throughout western Christendom in the wake of the first explorations of the Americas (Grafton 1992). It is against the backdrop of the New World discoveries that we will now turn, in conclusion, to the encyclopedic compilations of insect life that appeared in the last years of the Renaissance.

Two Renaissance texts that reflect the impact of New World discoveries on the study of insects are Ulisse Aldrovandi's *De animalibus insectis* (On Insects), published in 1602; and Thomas Moffet's *Theater of Insects,* compiled in 1590 though not published until 1634, several decades after his death, by Theodore de Mayerne. Both men were renowned naturalists in their respective countries of Italy and England, and both broke new ground in the concerted attention they gave to insects as a unique class of creatures meriting further study and exploration. That is to say, both texts reflected a break from the medieval and early Renaissance custom of discussing a relatively small number of familiar and iconic insects, such as the locust or the fly or the beetle that we have already encountered in the vermin trials and Northern Renaissance art. The works of Aldrovandi and Moffet were intended instead as exhaustive catalogs of every insect known in their time. Their focus was not on this or that particular insect, but rather on insects *qua* insects as a subset of the animal world, important in its own right.

In his illuminating essay on these two naturalists, Hugh Raffles has analyzed their works as an extension and embellishment of Joris Hoefnagel's vision of insects in the *Four Elements*, and there are indeed critical similarities that deserve to be mentioned (Raffles 2010: 123–40). The first is the sheer delight that all three men shared for the diversity of insect life. As we have seen, Hoefnagel was interested in a class of animals emblematic of nature's sheer fecundity, just as Aldrovandi and Moffet were. The second similarity noted by Raffles relates to the showcasing or foregrounding of insect life as the primary focus of attention—away from the margins of texts and into their centers. Here we might credit Dürer himself as the pioneering artist who first took this step, but Hoefnagel extended this courtesy beyond the portrayal of a lone stag beetle, again prefiguring Aldrovandi's and Moffet's celebration of the diversity of insect life.

A third similarity between the Flemish artist and the Italian and English naturalists was their shared rejection of contemporary fear and loathing as directed against insects, commonly perceived as an inferior and alien life form.

Raffles points out that this rejection amounted to nothing less than a critique of Aristotle's hierarchical taxonomy of animals, in which creatures were ranked according to their level of perceived perfection. In Aristotle's writings, insects figured at the bottom of this hierarchy, given their perceived enmeshment with the putrefying earth, while at the top stood the human male. In Hoefnagel's art, however, insects' association with fire—considered to be a celestial element—actually rendered them as significantly higher up on the teleological chain of being, thus justifying serious study of their habits, form, and nature.

Further uniting Hoefnagel with the later proto-naturalists is an overwhelming sense of wonder that all three men seemed to share in the face of the insect world. This aesthetic awe mirrors the religious mode of contemplation, intensified in *Four Elements* by the juxtaposition of insects with Pedro Gonzalez, the hirsute man. Of the two naturalists, it is Moffet who most explicitly returns to the theological importance of such wonder in his writings, explaining in the preface to his *Theater of Insects* that "no nation nor society of men amongst whom there is any religion of the gods, have any thing comparable to the Eleusian or Samothracian ceremonies, yet these do set forth but obscurely the Mind of All Things which they profess, which is plain enough in the fabric of all living creatures" (Moffet 1658: 3). The study of God's creatures, in other words, is a royal road to devotion to the Creator, with insects figuring for Moffet as the most wondrous pathway of all.

And why is that? The answer goes straight to the heart of many modern analyses of religious emotion: the wondrous is that which evokes a coincidence of opposites, stirring up in the mind of the beholder a mixed reaction of attraction and consternation (Otto [1923] 1958). Consider the observation of Dürer's oeuvre that the truly beautiful may lie in ugly things, or Hoefnagel's inclusion of a civilized hirsute in the fire folio, or even Kircher's obsession with the filthiness of the scarab as a sign of Christ. For Moffet, insects are even more wondrous than other animals because of their (oftentimes) minute size, in light of which the exquisite designs of their forms is even more marvelous:

> And such creatures as you cannot possibly dissect, I will make you to admire the more, the smaller they are ... Farewell then all those that so much esteem of creatures that are very large. I acknowledge God appears in their magnitude, yet I see more of God in the history of lesser creatures. For here is more of prudence, sagacity, art, ingenuity, and of certain evident divine being.
>
> (Moffet 1658: 3)

There can be little doubt that the theological rationale of *Four Elements*, which preceded the volumes by Aldrovandi and Moffet, prefigured many of the underlying justifications that drove the proto-modern works of entomology.

And yet there is something unique about the naturalists' passion for insect life that sets it aside from the purely devotional motivation of the Flemish painter. As committed as they are to glorifying the works of God, Aldrovandi and Moffet are equally if not more so driven by a zeal to know the *breadth* of insect life, and by a fervor to search out, track down, sketch, bag, preserve, or otherwise secure every last known specimen of insect and then to present them, side by side and en masse in the pages of their encyclopedic works for the readers who might subsequently learn about them. There is a different mode of wonder reflected in these works than what we find in Hoefnagel, who was after all, not a collector of insects but simply a painter of them.

By the time Aldrovandi and Moffet set out to gather up and catalog the sum total of all knowledge containing insects, European society as a whole had been on a quest of discovery for nearly two hundred years. This search began in the fourteenth century as a hunt for new knowledge of the ancient Mediterranean world, and the first acquisitions were ideas in the form of texts. But beginning in the late fifteenth and early sixteenth centuries, in the wake of the Columbian voyages, this orientation towards the exotic both accelerated and turned in a different direction. This was towards the Atlantic world and its promises of new land, wealth, power, and knowledge.

Grafton's terminology of the "shock of discovery" refers to the gradual realization among Europeans that none of their sources of authoritative knowledge could account for the sheer facticity of the American continents with their alien flora and fauna, peoples, and languages. This does not mean, of course, that biblical or classical sources were suddenly jettisoned in the wake of discovery. Rather, it suggests that scholars in fields purporting to explain the observable world could no longer accept unquestionably the authority of tradition. Another mode of knowing was required, one that approached the world directly, through observation, and built its conclusions on the basis of reason. In other words, the new world discoveries played a critical role in catalyzing the modern scientific method; they were an impetus to empirical exploration and rational thought.

As panoptic surveys of the new and enlarged knowledge of insects in the wake of discovery, *De animalibus insectis* and *Theater of Insects* are precursors to scientific literature, rather than examples of rational empiricism per se. These collections resemble nothing so much as the sixteenth-century cabinets of curiosity, which also played an important role in the history of early science by presenting contemporaries with the "raw data" of exploration, which future researchers would subsequently examine more carefully. Precursors to modern museums, cabinets of curiosity rose to prominence among both academic and political elites as spaces where the learned and/or powerful could both store and display the physical objects of cutting-edge discoveries unfolding in both their own societies and in the newly discovered Americas. As a rule, they were eclectic and eccentrically organized, mixing together such things as Catholic relics, miniature paintings, animal bones, Native American artifacts and, oftentimes, utter fabrications, such as purported dragons' eggs or feathers claimed to come from the phoenix bird. In German, these quasi-museums were termed *wunderkammern*, or "wonder chambers," in a clear allusion to the new, this-worldly sense of wonder as including the allure of the exotic.

Cabinets of curiosity played an important role in both Aldrovandi's and Moffet's research. Aldrovandi created one of the largest wonder cabinets of his day, in Bologna. Featuring over eighteen thousand items, it included specimens of the natural world, mostly from Europe but a few from the Americas, including a western balsam tree from Peru, an alligator, and a bean from Brazil; it also housed a number of Aztec artifacts from Mexico (Arellano 2015: 8). Aldrovandi had hoped to one day organize an expedition to the Americas, but instead remained in Bologna where he taught botany, zoology, and geology (Appleby 2014: 99). Moffet, in contrast, did not own a wonder cabinet, but in choosing to include the word "theater" in the title of his work, suggested that his text was meant to function like one, since another term for a wonder cabinet was a "theater of nature" (Appleby 2014: 100). Like Aldrovandi, the English naturalist also took an active interest in the American discoveries. In 1587, he met John White, the watercolorist and one-time governor of Sir Walter Raleigh's failed Roanoke Island colony, located off the coast of present-day North Carolina. White, in fact, bequeathed to Moffet some of his watercolors of American insects, including portraits of the previously unknown firefly and the swallow-tail butterfly, both of which were included in the *Theater of Insects* (Moffet 1658).

The insect encyclopedias of both Aldrovandi and Moffet resembled wonder cabinets in at least three ways. First, they served as figurative storehouses of information consolidating data on their subject into a single source. Second, they made no attempt to distinguish fact from fiction; each individual entry simply gathered together all the available information on a given insect, including empirical descriptions, folkloric anecdotes, and commentaries by medieval or classical authorities. Third, the volumes as a whole made little attempt to draw out underlying connections between the entries or ascertain deeper patterns in the data. On the one hand, the compendia were modern in the sense that entries were organized alphabetically and not, as in the *Four Elements*, according to a particular metaphysical cosmology. On the other hand, they contained very little that could be called a taxonomy. The one exception to this last point was Moffet's suggestion, in his preface, that a particular insect embodied the whole of nature's design. This notion of the "all in the small" was in fact a taxonomic method derived from hermetic metaphysics, which Moffet's contemporary John Dee had popularized among England's educated classes.

In the final analysis, Aldrovandi's *On Insects* and Moffet's *Theater of Insects* were quintessential Renaissance texts in their mixing together of old and new styles, as well as assumptions about the natural world. As products of a society profoundly altered by the shock of discovery, they reflect the beginnings of a modern scientific entomology. At the same time, they are still grounded in a religious sense of wonder for the insect world. In subsequent centuries, this nature-based theological orientation would continue, in a slightly altered fashion, in the form of Deism—the argument for God's existence inferred from the intelligent design of the universe. As for all the other features of these works, these would fade away into the mists of the past, together with lawsuits against cockroaches, hieroglyphic monads, and the feathers of the phoenix bird.

CHAPTER SIX

Insects as Symbols

Symbolism in the Christian Renaissance

LUCINDA COLE

In his *Pharmacopoeia Londinensis, or The London Dispensatory* (1653), the physician, botanist, and astrologer Nicholas Culpeper instructed Physicians in the Royal College of London to look closely at the insects living on "oak apples," the galls that form on oak trees, for knowledge of the future. If these withering galls "be broken" before a hole has formed, he writes, "one Living-creature or another" may be found in them, each signifying a different ecological future. Here he alludes to *De Materia Medica of Dioscordas* (1544), a foundational herbal written by the Italian naturalist Pietro Andrea Mattioli (1501–1577), also known as Mathiolus.

> *Mathiolus* saith if Oak Apples be broken asunder about the time of their withering before they have a hole through them, they contain in them one Living creature or another, which if it be a Fly it signifies War, if a Spider, Pestilence, if a Magget, Murren of beasts, if a Worm, scarcity of Victuals, if an Ant, plenty of Corn for us; I'le bind no bodies faith to believe it, because I never tried it myself; this I say, they are try and binding: being boyled in milk and drunk, they stop fluxes and the terms; and being boyled in Vineger, and the body anointed with the Vineger cures the Itch.
>
> (Culpeper 1653: 52)

Culpeper's goal in *The London Dispensatory* is to update knowledge about plants, animals, and medicines in order to advance the cause of medicine. In this carefully worded passage, Culpeper forwards Mattioli's symbolic interpretations of insects as prophetic agents, but balances them with an appeal to the value of observational experience, if not exactly experiment, by remaining agnostic about symbolic interpretations ("I'll bind no bodies faith to believe it") and by promoting his firsthand knowledge about insect galls, those "try and binding." His balancing act is in many ways typical of late medieval and early Renaissance writing about insects because he sees insects existing in several complex and overlapping semiotic systems, some emphasizing their symbolic meanings—classical, biblical, and popular—and others stressing their use value and morphology. In Culpeper's view, flies, spiders, maggots, worms, and ants in galls *could* in fact portend war, disease, famine, or bountiful harvests, but these associations need to be tested by observation and confirmed by experience.

Rather than dispel traditional symbolic interpretations, then, Culpeper opens them up to external means of verification in a way that reinforces Brian Ogilvie's view of how naturalists treat insects in this period. As Ogilvie argues, natural philosophers who now are relegated to the "prehistory" of entomology—Ulisse Aldrovandi, Johannes Goedaert, Thomas Moffett, and Jan Swammerdam, among others—sought to create systems of order in which "the emblematic meanings that insects vouchsafed humanity" co-existed with empirical understanding (2008a: 16). Empirical methods do not lead straightforwardly to the rejection of traditional beliefs (Stolzenberg 2013: 30). Traditional beliefs, in fact, could be used to shore up empirical methods.

From this perspective, the symbolic meanings of particular insects are created through and reside in complex networks of often occult associations, most impossible to trace to a single source. Many affiliations between particular insects and their significations were deep-seated in folklore and popular culture through connections now lost to us. In linking the maggot with famine, the spider with pestilence, and the fly with war, for example, Mattioli repeats what one historian of insects calls "a very common and old superstition," which, as late as the mid-seventeenth century, Sir Thomas Browne was still trying to understand. In his *Pseudodoxia Epidemica*, or *Vulgar Errors* (1672), Browne registers his skepticism about the prognosticating oak-apple because, he asserts, spiders are rarely found with maggots and flies on the same trees. At the same time, however, he acknowledges a kind of analogical truth grounded in literary and folkloric understandings. Homer, he notes, "compares the valiant *Grecian* unto a fly," then shifts to arguing that some spiders "are of a very venomous Nature," and some maggots "destroy the fruits of the Earth," heralding the possibility of famine (Browne [1672] 2009: 81). What endows the gall with prognosticatory power is an act of imagination he calls "Emblematical phansie": a well-known property of an insect is linked to a natural or literary "truth," and yet such truths have to rely on inherited chains of associative logic.

The insect-ridden oak apple serves as kind of cautionary tale, reminding us that chasing down the "meaning" of particular insects can be a fool's game. Nevertheless, certain insects enjoyed, or suffered, a fairly consistent place within a literary culture defined by classical literature, the Bible, and emblem books. The King James Bible contains over 120 references to insects and arthropods: ants, bees, fleas, flies, grasshoppers, grubs, hoppers, lice, locusts, maggots, midges, mole crickets, moths, scorpions, spiders, and worms, along with some identified as "swarming things" (Kritsky 1997: 183). Although both the nature of these insects and their specific meanings were sometimes hotly contested, biblical passages served as rich repositories of cultural associations and practices.

Equally important were the emblem books to which Browne alludes. Published widely in Britain and Continental Europe in the early modern period, emblem books drew on Greek, Roman, Hebrew, Christian, and Egyptian history, providing readers with a trove of biblical allusions and polyglot associations. Peter Daly emphasizes the founding assumption of emblem books—that objects throughout history and over time can be "read" in terms of originary qualities or attributes—if

> the cosmos [is] a system of correspondences and analogies, in which each object carries meaning imprinted into its very qualities by God at creation, then the interpretation of reality—that is the meaning read out of individual objects—is … a recognition of inherent meaning.
>
> (2014: 59)

Like the Bible, emblem books often featured insects, but, appealing to both the mind and the eye, they attempted to provide multisensory access to truths. Images and texts worked together as a kind of visual theology, making spiritual truths accessible to common people.

Because it is impossible to generalize about the meaning of every insect in the period 1300–1600, this essay focuses on several representative examples of how symbolic insects functioned as agents in larger cultural and semiotic systems. The first section of this essay examines associations between witches and moths. About the same time as Culpeper was instructing London's physicians, Joseph Glanvill, whose *Plus Ultra or the Progress and Advancement of Knowledge Since the Days of Aristotle* (1668) defended the experimental practices of the Royal Society, described cases of demonic possession in which insects, as the familiars of witches, played crucial roles. The trial of the accused witch Elizabeth Style demonstrates how the symbolic meanings of insects could have fatal consequences. The following sections examine three mutually reinforcing discourses in which insect symbolism helped normalize competing notions of self and other and the natural world. The second section explores how insects were deployed in religious and sociopolitical controversies in early modern Britain. For several centuries, English Catholics and Protestants used insects symbolically as political and theological weapons. Flies, locusts, and spiders came to stand in for ecclesiastical heretics and their corrupting offspring. This weaponizing of insects drew on the Bible as well as on a complex and transcultural emblematic and fabular tradition, the subject of the next section. In the emblem tradition, insects—especially bees—are powerful symbolic tools for visualizing moral precepts and a range of social and religious values, including concrete notions about what constitutes the ideal society, and the roles of princes and subjects within it. The final section examines the dung beetle, and theories of generation in the works of the Renaissance polymath Athanasius Kircher. In Kircher's syncretic and idiosyncratic method of interpreting the dung beetle across several religious traditions, we glimpse how new scientific texts coexisted with defenses of a learned philological tradition. Throughout the Renaissance, insects helped to shape the symbolic economies of traditional and scientific knowledge systems, bridging theology, politics, economics, and history.

FAMILIAR FIGURES: WITCHES AND INSECTS

Cats, toads, rats, weasels, and other small mammals and amphibians have a long history as the most common familiars for witches in European and other cultures, ancient and contemporary. The ancient Babylonian deity Beelzebub was worshipped in the form of a fly; in Norse mythology, Loki, the god of mischief and destruction, also assumed the shape of a fly. Talking spirits take the form of house crickets in Malay, and horse-flies that enter some Native American villages were thought to carry the souls of dead relatives and to portend destruction (Weiss 1930). Jean Bodin, who wrote a particularly influential account of witchcraft and sorcery, claimed that God gave Satan power over flies and other "less exalted" or "unclean" animals (Conner 2006: 37; Cole 2016: 24–48). This power, in turn, could be borrowed by witches or demons who frequently assumed the forms of insects to harm the property and bodies of witches' enemies, including ecclesiastical and legal authorities.

In their fifteenth-century *Malleus Malificarum*, Kramer and Sprenger argued that demons and witches cannot assume "natural" shapes except in the case of small and imperfect creatures, primarily insects, worms, frogs, rats, mice, and reptiles (1486: 109). The term "imperfect creatures" derives from Aristotle's distinction between "perfect creatures," born from sexual reproduction, and insects and vermin thought to generate spontaneously from putrefying carcasses or decaying organic material, like the slime and mud of the Nile. Imperfect creatures existed in a shadowy and deeply disputed space in both theological and scientific literature because early modern writers wrestled with the question of how a perfect God could have created insects and other creatures seemingly absent from the Garden of Eden. Alexander Ross puts the question this way: "Did God create in the beginning, imperfect creatures, as Bees, Wasps, and such like?" (1626: 18). To justify the ways of God to man, Ross relies on a broadly Aristotelian explanation: "He did not create them actually, as he did the perfect creatures, but hee created them in their first causes, as hee gave that faculty to the flesh of an Horse, to beget Waspes being dead" (18). In other words, although insects and other imperfect creatures were not part of his original creation, God made it possible for them to breed through secondary causes: the processes of putrefaction. Demons and witches, in turn, took advantage of mud, slime, and rotting animal carcasses. Unable to "create" their imperfect familiars, they could nevertheless "produce" them, according to this account in the *Compendium Maleficarum* (1608):

> It is usually the demon who, in accordance with his pact with the witch, produces such animals by the application of active to passive forces: for the witches themselves for the most part do not know how they are produced and are ignorant of the causes: as when the devil gives the witch a little dust which she throws into the air, and there are born various kinds of locusts and grasshoppers and mice and caterpillars and such like animals. The devil could also, having produced such things, tend them and nourish them and give them breath in remote places where they appear to be generated; as among rocks, where imperfect creatures are often born, such as flies and mice and similar things.
>
> (Weiss 1930: 131)

This explanation preserves the creative power of God, makes witches the passive pawns of demons, and casts insects as both vessel and symbol of the demons' presence and destructive power.

The symbolic, evidentiary power of insects as familiars is apparent in surviving accounts of witchcraft trials. The labourer Joan Wayte "did entertain, employ, and feed an evil spirit called a butterfly"; Alicia Warner de Rushmere employed "evill sperits" to "carry lice"; and Susanna Smith de Rushmere said the devil appeared to her in the form of a black bee (Weiss 1930: 131). In his *Saducismus Triumphatus* (1681), a collection of folklore about witches and witchcraft, the latitudinarian clergyman Joseph Glanvill reported on the trial of Elizabeth Style of Bayford and the symbolic role that insects played in her conviction and execution. Glanvill's account is worth examining at length because it reveals how the appearance of insects called forth a range of theological values and assumptions that identified them as symbols of demonic forces.

In her confession, Style admits to torturing Elizabeth Hill and to consorting with the Devil for ten years. Her familiar "*Robin*" appeared alternately in "the likeness of a Dog, and Cat, and a Fly like a Millar, in which last he usually sucks in the Poll [scalp or head] about four of the Clock in the Morning" (Glanvill 1681: 137). After her confession, she was put in a room and watched by two men. According to their account, "there came from her Head a glistering bright Fly" which vanished up the chimney; fifteen minutes later, two smaller flies emerged which "seemed to strike" at the hand of one of the men. At this point, perceiving a "change" in Style's "countenance," which became "very black and gastly," the examiner suspected that "her Familiar was then about her" (144–5). Seeing "her Hair shake very strangely," he picked it up, "and then a Fly like a great Millar flew out from the place, and pitched on the Table-board, and then vanished away" (144–5). Insect infestations are taken as evidence of demons partly under Style's power.

The legal evidence of Style's guilt, then, rested as much on the examiners' descriptions of the insects around her as it did on her neighbors' testimony and her own confession. Twice her familiar is described as being like a "millar," a white or gray insect or a moth; the moth, in turn, signifies decay, destruction, or death (Neri 2011: xi). In the Book of Job, "Death is a Moth, and as the Moth breeds out of the Garment, so Death do's out of the Body" (Allestree 1695: 6). In Style's case, the moth supposedly emerged from a sore, "like 'very red and like raw Beef,'" on her scalp (Glanvill 1681: 144–5). The fact that the flies "vanish" was a crucial piece of evidence. Witch-hunters were taught to confine suspected witches to a closed room and to watch for spiders or flies; if the insects could not be caught and killed, they could be regarded as the witch's "imps" (Weiss 1930: 132). Asked what "it was that went out of her Poll," Style at first "said it was a Butterfly, and askt them why they had not caught it" (145). When one of the men responded, "they could not," she teases them with her own knowledge of the symbolic roles that insects play in witchcraft: "I think so too" (145). When they again looked under her hair and found that the raw and bloody scalp had returned to its "former colour," she admitted that the "Fly" was in fact her "Familiar," coming to her (and from her) at its usual time (145). Style was found guilty of witchcraft and executed in 1663.

Elizabeth Style's case reveals a good deal about the nature of symbols themselves: symbols gains meaning only through the complex networks of associations and values in which they are embedded. Because these networks are themselves local, unstable, often contested, and interdependent, few insects symbols are genuinely transcultural and transhistorical. Within the discourse of witchcraft, for example, butterflies—a classical image of *psyche* or soul—are not the glorious and colorful harbingers of spring, but mischievous and even ominous scavengers who share with witches the habit or stealing butter, as their name suggests. The symbolic meanings of moths, spiders, butterflies, bees, then, ultimately cannot be divorced from the ecological, economic, and political contexts in which they exist.

ALLEGORY, LOCUSTS, THE ENTOMOLOGY OF THE END TIMES

In Renaissance theological literature, insects appeared most frequently in one of two ways: as mutable vessels for the Lord of Flies to wreak havoc on God's creation or as divine emissaries. In this regard, they can function either as symbols of a corrupt natural world, or signs of God's displeasure with a sinful population. In his 1616 treatise, *The Fall of Man, or the Corruption of Nature*, Godfrey Goodman exhorts his readers, "let not the plagues of Aegypt seem so incredible" when Jacobean England suffers its own plagues that are evidence of its corrupt nature. "Wee stand not onely in feare of fierce Lions, cruell Tigers, rauening Wolues, deuouring Beares," writes Goodman, "but Gats, Flies, and the least wormes doe serue to molest vs" (1616: 219). As the title of his treatise suggests, the fall of man is bound inextricably to the corruption of a natural world, figured not only by traditional images of wild beasts (all but extinct in seventeenth-century Britain) but by small, noxious and ubiquitous insects.

As I have argued elsewhere, Europe indeed had its share of animal plagues, even occasional plagues of grasshoppers or locusts (Cole 2016: 49–80). In his posthumously published *Theater of Insects*, Thomas Moffet (or Muffet) reports on insect-induced famines in France in 455, 874, 1337, 1353, and 1374; in the last of these, a third of the inhabitants died (Topsell 1658: 987). In 1476 locusts "wasted almost all *Polonia*"; in 1536 they invaded first eastern Europe, then Germany and Italy, before returning in 1543, when their bodies "form[ed] heaps above a cubit high" (Topsell 1658: 987). At times during the early modern period, locust-like insects devastated crops as far north as Wales and Ireland (Cole 2019). While Moffet was drafting his chapter on locusts, he "received news that the Spaniards were sorely afflicted with swarms of Locusts brought thither out of Africa":

> For they flew like Armies through the skies and darkned [sic] the air. And the people when they saw them, rang all their bels, shot off ordinance, sounded with trumpets, tinkled with brazen vessels, cast up sand, did all they could to drive them away; but they could not obtain what they desired, wherefore sparing their labour in vain, they died everywhere of hunger and contagion: as the Mariners and steer-men reported to us, who escaped very hardly from that danger themselves.
>
> (Topsell 1658: 987)

This military imagery renders locust infestations as dire threats rather than merely theological or political symbols. The Spaniards, Moffet reports, respond with the weapons and signals of war in trying to repel these "Armies ... that darkned the air." Locust swarms can extend from 150 to 400 square miles; fifty million locusts can eat one hundred tons of food every night (Schmetzer 1988). Even if England—for the most part—offered a less habitable climate to these hungry insects than southern European nations, the English were vicarious witnesses to the damage they could do.

Christian writers between 1400 and 1700, then, could look to the physical world and the historical record as well as to the Bible in order to situate insects within a symbolic cosmology of good and evil. In the theological tradition that writers like Moffet inherited, the Four Horsemen of the Apocalypse—War, Famine, Disease, and Death—were often heralded by, if not directly attributed to, insect infestations, "given power," according to Revelations 6:7–8, to "kill with sword, and with hunger, and with death." The Puritan writer George Wither in *Britain's Remembrancer* (1628) begins his poem with a description of how Famine, personified as a general, commands a host of "Troups" that include caterpillars, locusts, birds, and worms:

> The crawling *Caterpillars*, wastfull *Flyes,*
> The skipping *Locust* (that in winter dies)
> *Floods, Frosts, & Mildewes, Blastings, Windes, & Stormes,*
> *Droungh, ravnous Fowles, & Vermine, Weedes, & Wormes:*
> *Sloth, Evill husbandry*, and such as those,
> Which make a scarcenesse where most plenty grows.
>
> (Wither 1628: 41)

Although Wither puts locusts under the command of Famine, in the Bible they usually are represented as part of God's militia, as "war horses" or a "great army" (Joel 2:4–5). Naturalists tended to adapt the symbolic biblical description of insects as part of God's army for their own accounts. Locusts, according to Moffet, are "amongst the Armies of the Lord of hosts, when he pleaseth to punish the sins of men, and to revenge himself on the despisers of his Lawes" (Topsell 1658: 987). In his *Theatre of Politicall Flying Insects* (1657), Samuel Purchas uses almost identical language: we "must conclude that these small creatures have a chief place among Gods Troops and Armies" (1657: 201). The threat of verminous scourges structures biblical and natural history narratives alike, and insects in early modern texts play a crucial role as agents of divine displeasure.

Locusts, significantly, are similarly active agents in universal histories, where the lines between history and political and religious polemics are seldom clear (Markley 1999). John Pory, for example, a protégé of the geographer Richard Hakluyt, is best known for his translation of English translation of John Leo Africanus's *A Geographical Historie of Africa* (1600). In Africanus's text, locusts are the last species on a list of animals not generally found in Europe; they come in "[M]onstrous swarmes," devouring "trees, leaves, fruites, and all greene things growing out the earth," sometimes "procuring ... extreme dearth of corne" (Africanus 1600: 349). Although the biblical overtones of the eighth plague of Egypt hover over this passage, Africanus mitigates the disaster by noting the adaptive measures taken by indigenous peoples; he describes how the people of Arabia and Libya "esteeme the comming of these locusts as a fortunate boading: for seething or drying them in the sun, they bruise them to powder, and so eate them" (349). Because the locusts in Africanus lead to celebration, rather than conversion, Pory steps out of his role as translator of Africanus to include another, very different account of locust infestation that demonstrates, he writes, how God uses locusts "as a most sharp scourge between times to discipline all the nations of Africa" (958). The supplementary description is by the Christian writer Paulus Orosius, a contemporary of St. Augustine's, whose *Seven Books of History Against the Pagans* was frequently widely excerpted during the Renaissance. In Pory's excerpt, Paulus tells the story of "an horrible and extraordinarie destruction" visited on Africa; God sent locusts that devoured corn, bark, and even the wood of the trees. Although a "violent and sudden winde" bore them aloft and out to sea, their "lothsome and putrified carcases" were cast back onto shore, where "an incredible stinking & infectious smell" brought about a general pestilence, which first killed birds and beasts, then men by the thousands (350). "Never [before] in the time of Christians," Paulus Orosius concludes, did a "scourge of locusts" bring such devastation (351). By promoting this polemical account of infestation over Africanus's more neutral one, Pory transforms geography into biopolitics. Locusts set the stage for a new physico-theological order: Christians were given license to rid Africa and India, figuratively and literally, of insects that marked "pagan" difference from Christendom and its divinely sanctioned regimes.

As these cases suggest, early modern writers often walked a fine line between naturalistic and typological readings grounded in the Bible. Over the next several centuries, extending well beyond the Renaissance, an industry of biblical interpreters and interpretations emerged to try to stabilize these symbolic meanings for a general populace. Among the first of these was Thomas Wilson, an Anglican minister at Canterbury, whose *Christian Dictionarie*, according to its subtitle, explained "the signification of the chiefe words" of the Old and New Testaments in order "to increase Christian knowledge." Most of these words are conceptual ("blessed," "dead," "guilt," "work of days"); several referred to birds and quadrupeds ("eagle," "wolfe," "swine," "dragon," "sheep," "foxes," "dogs"); and several to insects ("spider," "scorpion," "locust"). His entry on locusts follows his characteristic pattern, dividing the definition into two parts, the first grounded in natural history:

> A certaine vile Creature so called. There were diuers kindes of them; some hurtful & venomous; others, commodious for meat. *Mat.* 3, 4. *His meat was Locusts.*
>
> (Wilson 1612)

Wilson alludes to the double ecological role of locusts, both as destroyers of crops and as a source of food. To illustrate the latter, he cites the story of John the Baptist wandering through the desert, surviving on locust meat and wild honey. The second definition, on the theological and symbolic signification of locusts, in contrast, is cast in terms of England's civil strife:

> Deceiuers, or authors of false Doctrines, generally: and more specially; the Popish Cleargie, with their swarmes of Monkes, Friars, Priests, and *Iesuites. Reu.* 9, 3. *There came Locusts out of the smoake.*
>
> (Wilson 1612)

His citation from Revelations, the most overtly symbolic book in the New Testament, is revealing. In Chapter Nine of Revelations, the "fifth trumpet" initiates a five-month period of torment on unbelievers:

> And out of the smoke came forth locusts upon the earth; and power was given them, as the scorpions of the earth have power. And it was commanded them that they should not hurt the grass of the earth, nor any green things, nor any tree, but only the men who do not have the seal of God upon their foreheads.
>
> (Rev. 9.3–4)

In the *Dictionarie*, Wilson equates Revelations' tormenting locusts with human sects: crop-destroying insects bear a metaphorical relation to "swarmes" of "Monkes" and "Friars" promoting "false Doctrines." For him, "Popish Cleargie" signify the corruption that heralds the end times.

The *Christian Dictionarie*, adapted and supplemented by later writers, ran through six editions. By demonizing and dehumanizing Catholics as locusts and scorpions, as deadly threats to spiritual, political, and natural order, Wilson helps naturalize anti-Catholic sentiment in ways that help reinforce political identifications and disidentifications during the next century. Insofar as Catholics are associated with "Swarmes," they can be marked as dangerous and unpredictable populations of noxious creatures. In anti-Catholic political discourse, the locust is less a stable symbol than it is an effective means of producing and reaffirming sociopolitical or theological affiliations, experienced or felt as much as they are consciously understood.

VISUAL THEOLOGY: EMBLEMS, STINGERS, AND BEES

Eric Jorink has argued that the "influence of the classical and biblical corpus on the appreciation of insects can hardly be overestimated" (2010: 187). The bee, the ant, and the butterfly, in particular, were regarded "primarily as symbols," respectively, of the ideal society, of perseverance, and of transience or resurrection (188). A crucial source for the dissemination of these symbolic meanings was the emblem book. Emblem books, both secular and religious, contained illustrations accompanied by texts, either proverbs or poems (Freeman 1948). "The canonical emblem," according to Mara Wade, "can be defined as a tripartite composition consisting of a brief motto in Latin or a European vernacular language (*inscriptio*), an enigmatic picture (*pictura*), and an epigram (*subscriptio*)" (Wade n.d.). Insects, when they are included in emblem books, function as overdetermined objects of contemplation to promote moral and religious knowledge. Their folk, historical, and theological associations become complex symbols within the "correspondences and analogies" that structure early modern thought.

Although in most emblem books insects play smaller roles than do charismatic megafauna such as deer and dogs, they often work as images that remind readers of the vulnerabilities of the social and natural orders. A popular and comparatively straightforward emblem, based on Aesop's Fables, is that of the Eagle and the Beetle; the lowly beetle takes revenge on an arrogant eagle by boring into and destroying her eggs. Andrea Alciato included it in his enormously influential *Emblematum Liber* (1534), and the image was adopted with little variation in subsequent German, French, and Italian texts, often with the inscription *A mimumus temiendum* ("Even the least are to be feared") (Alciato 1534). Another noteworthy emblem in Alciato features a hand holding a cricket by one wing, while the other hand chases away surrounding flies. It is entitled "Against Detractors" and asks the question "What use is it to banish insects with busy flapping? More satisfactory to ignore what you can't destroy" (Alciato 1534). In several emblem books, flies or other flying insects hover around candles, representing temptation, danger, foolishness, or the inevitability of death. The fly is the only image of an insect in Frances Quarles's *Emblemes* (1635), which went through three editions in his lifetime. In the preface, Quarles begins by asserting: "Before the knowledge of letters God was knowne by *Heiroglyphicks*: And, indeed, what are the Heavens, the Earth, nay every Creature, *Hieroglyphicks* and *Emblemes* of his Glory?" In one edition,

Hieroglyph V depicts an angel shielding the light of a candle from a blast of air, while a winged creature hovers dangerously close to the flame on the other side of the shield. In many of these instances, flies and other insects invite the reader to imagine the transitory nature and insignificance of this life in comparison with eternity.

Scorpions and spiders, which are not technically insects at all, frequently appear in emblem books as representations of treachery or punishment. Cesare Ripa includes in his *Iconologia, or, Moral Emblems* a figure of Malice as an "old Hag" covered with spiders, "which show that *Malice* is like them, which weave deceitful Nets for Flies" [1603] 1709: 55). This emblem draws on a long tradition of medieval and early modern art in which spiders are deployed as symbols of treachery; Judas often appears as a spider or scorpion, and Christ frequently tramples a scorpion under his feet. During the religious controversies sparked by the Reformation, the symbolic interpretations of these emblems were turned into tests of theological and political allegiance. John Heywood's long, illustrated poem *The spider and the flie* (1556) features a Protestant spider and a Catholic fly, caught in the spider's web, debating whether Catholics have a right to exist. Both the spider and fly enlist the aid of other insects—ants support the Protestant spider and butterflies the trapped Catholic fly—in a battle that kills five thousand of each species. Because Heywood was a devoted Catholic, however, the housewife Mary (Mary Queen of Scots) eventually crushes the Protestant spider. Heywood's symbolism captures the ways that seemingly straightforward allegory could marshal the physical qualities of insects to promote specific political and religious agendas.

A number of emblems feature a spider at the center of a web, but they frame the moral purpose of this image in different ways: as helplessness, perseverance, or legal power. As these varied significations suggest, although the emblem tradition tries to conjure up stable, unitary meanings for objects and animals to situate their essential attributes within the divine order of things, the very acts of representation produce different assumptions about what symbolic meanings spiders, flies, and beetles embody. In the case of spiders and insects, many sixteenth- and seventeenth-century emblem books demonstrate that authorial ingenuity can overwhelm ostensible commitments to the revelation of stable symbolic meanings.

The most important exception to this interpretive free-for-all, as generations of scholarship attests, is the consistent symbolic role that bees play in emblem books. In his *Natural History*, Pliny the Elder argued that bees are the only insect seemingly made for the beneficial use of man (Pliny 1855). Materially, of course, bees make honey; symbolically, their hives provide an idealized model of an ordered and hierarchical society. Medieval bestiaries emphasize bees' ability to work tirelessly, live in a harmoniously structured community, and choose their "king" or, as it was discovered in the late sixteenth century, their queen

(Butler 1609). St. Anthony of Padua emphasizes the role of Christ as leader of a hive whose inhabitants—penitents seeking salvation—must follow "like good bees, and to fly with Him, and to place Him, that is His faith, in ... our hearts" (Neale 1856: 244). Building on this symbolic identification of bees and Christian virtue, later moral authors and designers of emblem books emphasize the "natural" division of the labor in the industrious and well-ordered beehive: some bees fight off intruders, some build the hives, some collect honey, and some watch for rain. "And yet," writes one thirteenth-century commentator, "among so diverse works none of them doth espy or wait to take out of other's travail, neither taketh wrongfully, neither stealeth meat, but each seeketh and gathereth by his own flight and travail among herbs and flowers that are good and convenable" (Steele 1803: 175). In the early seventeenth century, Moffet emphasized the special status of bees in the symbology of the natural world: "Whereas the Almighty hath created all things for the use and the service of Man, so especially the bees": bees are "patterns and presidents of political and oeconomical vertues" as well as "Teachers and School-masters" who can "instruct us in certain divine knowledge" (Topsell 1658: 644). In their positive and even spiritual symbolic roles, bees largely escape the interpretive fate of other insects, Aristotle's "imperfect creatures."

Emblem books provide numerous illustrations of Moffet's emphasis on the political and economic virtues of bees as a form of exemplary moral instruction. One of the most persistent apian metaphors, spanning classical and Renaissance literature, is the image of the stingless king bee, or the king who has a stinger he never uses. Because "he" does not attack other bees in the hive, "he" becomes the model for a king who "balances justice with mercy, charity with clemency" (Woolfson 2010: 288). In Alciato's Image 148, "Clemency in a Ruler" (Figure 6.1), insects that look like bees (but are identified as wasps) fly around a manmade hive; a larger version of the same species hovers at the top, and his role is explicated by this inscription: "The king of wasps will never implant any sting, and will be twice as big as the rest. This be a sign of mild dominion, a disciplined kingdom, and inviolable law entrusted to good judges" (Tung 1621). Alciato's emblem pictures an ideal kingship that rests on unquestioned authority but that never descends to violence and never violates "sacred" laws.

Geoffrey Whitney's *Choice of Emblemes* (1586), the first English emblem book in the style of Alciato, borrows this (and eighty-six other images), explicitly connecting this "maister bee" to Alciato's stingless wasp: "Hee hath no sting, yet none can doe him harme, / For with their strengthe, the rest about him swarme" (Whitney 1586: 200–1). Whitney expands Alciato's short poem on mercy into an eight-stanza reflection on Richard Cotton's estate, Combermere, originally an abbey, later remodeled, after the Reformation, as a country house. In this instance, a traditional emblem about benevolent kingship shades into

178 ANDREAE ALCIATI

Principis clementia.

EMBLEMA CXLVIII.

VESPARVM quòd nulla vnquã Rex spicula fig
Quódque aliis duplo corpore maior erit;
Arguet imperium clemens: moderatáque regna,
Sanctáque iudicibus credita iura bonis.

FIGURE 6.1 Alciato Image 148, "Clemency in a Ruler," http://www.emblems.arts.gla.ac.uk/alciato/dual.php?id1=A91a148&type1=1&id2=A15a149&type2=1.

an English country house poem, even a defense of the new Protestant order. Whitney commends the "stately seat," its fish, fowl, and "cattle sundry flocks," its "crystal springs" and compares the estate to a hive: "This is the hive, your tenants, are the bees— / And in the same, have places by degrees" (1586: 25–31). Comembere's conversion from a monastery to a private estate managed by bees is a timeless image of a productive social order. Moffet's "political" and "oeconomick vertues"—the ruler atop a laboring and thrifty society—become bound together in the image of the hive (Figure 6.2).

The sociopolitical significance of the bee as symbol structures one of the more intriguing cases of scientific bias in the history of natural philosophy: the widespread and gendered assumption that, as an ideal society, the hive had to be ruled by a male, who arose *ex nihilo* from rotting carcasses or mud. Despite ongoing debates in the early modern period and some (often disputed) evidence that hive "rulers" laid eggs, this belief in masculine authority among bees was not discredited until Charles Butler, a beekeeper, published *The Feminine Monarchie, or the Historie of Bees* in 1623. Butler's natural history

FIGURE 6.2 Whitney Patria cuique chara, http://www.mun.ca/alciato/whit/w200.html.

challenged directly the values and assumptions of a masculinist ecclesiastical authority and the symbolism on which it rested. In the Catholic tradition, the church often was figured as a masculine beehive with a universal ruler—the Pope; in contrast, Protestants, like John Calvin, who dismissed this idea of a universal papacy, used the beehive as an image of individual churches or dioceses, with their own bishops (Woolfson 2010: 292). Both church and state throughout much of the early modern period were invested in structures that supported different authoritarian versions of masculine rule; the emblem tradition afforded them a symbolic model and justification even after natural histories of bees no longer did.

To some extent, the values and assumptions about flies, spiders, and bees in emblem books shape even the practices and discourses of natural history. Early modern emblems and natural histories shared similar sources and similar purposes: to reveal the operations of divine order. Both sought to go beyond words—the conceits of human wit—and to ground knowledge in material reality. Aldrovandi's *De Animalibus Insectis* (1602) included a chapter on bees, discussing them in relation to their historical, moral, mystical, emblematic, proverbial, fabular, and iconic meanings. Ogilvie notes that Aldrovandi collected symbolic meanings without insisting on the primacy of one over another: "he did not systematically develop the moral or ethical implications of insects" (2008a: 9). In contrast, Joachim Camerarius's *Symbolorum et Emblematum*, which focuses exclusively on flora and fauna, is organized according to classificatory schema found in natural histories (Smith 2017). His volume on birds ("flying animals") and insects, *Symbolorum et Emblematum ex volatilibus et insectis* (1596), begins with the eagle and ends with the phoenix; this structure mirrors that of Conrad Gesner's *Icones Avium Omnium* (1550), which nonetheless opens with an "extended discussion of animal symbolism" (Kusukawa 2010: 306). To read emblem books and natural history together is to appreciate how knowledge of insects was mediated by a network of visual images inherited, adapted, and circulated over centuries.

With the rise of microscopy, insects further became identified with the argument from design: the wisdom and power of God could be deduced from the beauty and utility of all aspects of his creation (Markley 1993). Johannes Swammerdam and Antoni van Leeuwenhoek claimed that the anatomical structure of insects constituted proof of God's handiwork, "no less to be admired," writes one Dutch naturalist, "in a mosquito, a fly, a flea, or a cheese mite than in the creation of the largest elephant" (qtd. in Jorink 2010: 182–3). The English naturalist John Ray echoed this argument in his *The Wisdom of God Manifested in the Works of the Creation* (1691), a crucial text in the physico-theology of the Royal Society. In the eighteenth century, a German book called *Insecto-theology: Or, a Demonstration of the Being and Perfections of God* updated this British tradition; its frontispiece is dominated by the *inscriptio* "*Maxima in minimus*" (in the least, most). Although emblem books were of less and less significance in the eighteenth century, the scientific study of the natural world was still shaped, in varying degrees, by the assumptions and values that they had promoted for centuries.

INSECTS, SOULS, AND COMPARATIVE RELIGIONS

I have argued that locusts, moths, spiders, and bees are critical players in the discursive, material, and symbolic systems of early modern Christian folklore, theology, and politics. Insect semiotics reflect a Christian hierarchy of value: bees embody positive moral, political, and socioeconomic virtues, honey for the soul, while most of the other insects appear either as nuisances that should be contained, if not controlled, or as active spiritual dangers. Yet early modern writers on insects confronted difficult theological questions of how and why an omniscient God would create imperfect creatures, excluded from the Garden of Eden and seemingly bereft of souls. Creeping and swarming things—other than bees and butterflies—hover noxiously within the human-dominated order of creation that both Catholics and Protestants saw as their duty to protect. This problematical status takes structural form in seventeenth-century discussions of Noah's Ark. Neither John Wilkins, one of the founding members of the Royal Society, in *Essays Toward a Real Character and a Philosophical Language* (1668), nor Athanasius Kircher, in his *Arca Noë* (1675) allocates any space on Noah's Ark to caterpillars, insects, or other imperfect creatures. For Kircher, insects were not brought on the Ark because they were generated in the dung of the other animals; for Wilkins, rats, mice, moles, and insects did not require any plotted space because they found "sufficient room" in the "Stalls" of other animals (Wilkins 1668: 165). As Wilkins' comments suggest, insects as a group either had to be explained away in Judeo-Christian descriptions of the natural order or accommodated only on the margins of the biblical text.

This question about the difference between "perfect" and "imperfect" creatures was complicated further by European encounters with other cultures. By the second half of the seventeenth century, when Wilkins and Kircher wrote, defenders of the Judeo-Christian faith had to contend with theories that challenged human exceptionalism — the idea that humans are categorically different from other animals, including insects. A pillar of Hindu and Buddhist thought was the transmigration of souls, which circulated widely in Europe in the centuries after the Portuguese, Dutch, French, and British established trading posts on the Indian subcontinent. Even in Europe, some religious radicals revived the ancient doctrine of *anima mundi*, or the world soul, found in Pythagoras. A world soul implies intrinsic connections among all living, and often non-living, things, including insects and humans; it thereby calls into question the hierarchical models of creation favored by compilers of emblem books and natural philosophers alike. Not surprisingly, such doctrines provoked strident responses. The seventeenth-century physician and natural philosopher Thomas Willis, for example, derides the idea of a world soul as a "Platonick Fiction" (1683: 4). Although Willis, who did ground-breaking work in comparative neurology, admits that mammals *may* have a claim to something he calls a "soul," he flatly rejects the notion that insects do. He asks rhetorically:

> where do many Myriads of Souls, even innumerable, of Insects and Fishes, which are daily produced, subsist, and what do they? The Bodies of very many of these serve only as Food to other Creatures. And for that the Souls to these Bodies, serve chiefly to preserve them only for a little time, and as it were pickle them to keep them from putrefaction, there is no need that these should be therefore immaterial and immortal. Besides, when of old, Egypt was infected by Divine Punishment, with Swarms of Fleas, Flyes, and other Various Kinds of innumerable Insects, and that the same also abounded every where, it is not easily to be Conceived, from whence so many Souls were so suddenly Called, and into what places, the same being by and by separated, could be placed.
>
> (1683: 4)

Produced "daily," the insects serve largely as "Food to other Creatures," even as they multiply and swarm in their roles as instruments of "Divine Punishment." They are purely instrumental beings, called into existence beyond the regimes of human counting and calculation. Because their sheer number boggles the mind, it is impossible "to be Conceived" how their souls could pre- or post-exist their transitory lives. Countering the logic of Pythagorean metempsychosis, Willis argues that so many freed or "separated" insect souls could not be housed or "placed" in the more finite number of perfect creatures. Although insects may serve as a *metaphor* for the soul, as in the fly and candle, this relationship is *merely* figural. For Willis, the collapse of the figural constitutes an heretical drift toward granting insects souls.

A more syncretic approach to stabilizing Christian symbology is evident in the work of Athanasius Kircher, a contemporary of Descartes and Newton. Now more or less marginalized but in his day "one of Europe's most successful scholars," Kircher was also the last symbologist for whom the interpretation of the "hieroglyphs" of different cultures took precedence over the experimental method and even the argument from design (Stolzenberg 2013: 35). This polymath German Jesuit, "the last man who knew everything," poised on the edge of modern, discipline-specific knowledges, attempted to discover a universal means of communication derived from Coptic, an early Egyptian literary language that, he maintained, was derived from hieroglyphs (Findlen 2004). In some ways, Kircher's ambitions, if not his precise methods, recall the universalist and transcultural aspirations of the emblem book tradition. His *Oedipus Aegypticus* (1652–1654), an early and systematic attempt at comparative religion, purported to have solved the interpretive conundrums posed by Egyptian hieroglyphs. In Daniel Stolzenberg's view, this "book illuminates a complex moment when empiricism and esotericism coexisted," and although Kircher's readings of Coptic were based on "spurious documents," behind them "lay widely accepted principles about symbolic communications and the transmission of ancient knowledges" (Stolzenberg 2013: 6). Ancient knowledges could be treated as esoteric allegories with the potential to make Judeo-Christian interpretations compatible with Greek, Latin, Egyptian, Arabic, and Chinese religious and cultural traditions.

Much of *Oedpius Aegypticus* is based on a text called *Hieroglyphica*, which was discovered (or forged) in the fifteenth century and claimed to be an ancient Greek translation of mystical Egyptian texts (Figure 6.3). Kircher takes over from Horapollo, the supposed author of this treatise, a number of assumptions about Egyptian hieroglyphs, including that of the dung beetle. To depict the "only begotton, or generation, or a father, or the world, or a man, they delineate a SCARABAEUS," Horapollo claims, a creature "self produced" with no need of a female (1840: 20–1). "When a male is desirous of procreating, he takes the dung of an ox, and shapes it into a spherical form like the world"; he then rolls it "from east to west," shaping it in "the figure of the world" (21). The enterprising male scarab deposits the dung ball in a hole for twenty-eight days, until, at the conjunction of the sun and moon, animals "issue forth."

> The scarabaeus also symbolizes generation, for the reason before mentioned—and a father, because the scarabaeus is endgendered by a *father* only—and the *world*, because in its generation it is fashioned in the form of the world—and a *man*, because there is no female race among them.
>
> (21)

Horapollo's explanation of the scarab's symbolic power recalls the descriptions of the "king" bee derived from Pliny the Elder: it reinforces a masculinist theory of creation that dispenses with females entirely. The scarab's god-like powers derive from its mode of spontaneous generation.

Kircher takes pains to accommodate Horapollo's fanciful explanation to the Christian hermeneutical tradition, even as he promotes some of its unorthodox offshoots. As Ingrid Rowland argues, Kircher wanted to defend a doctrine of *panspermia*, the Paracelsian notion that the seeds of all things (*semina rerum*)

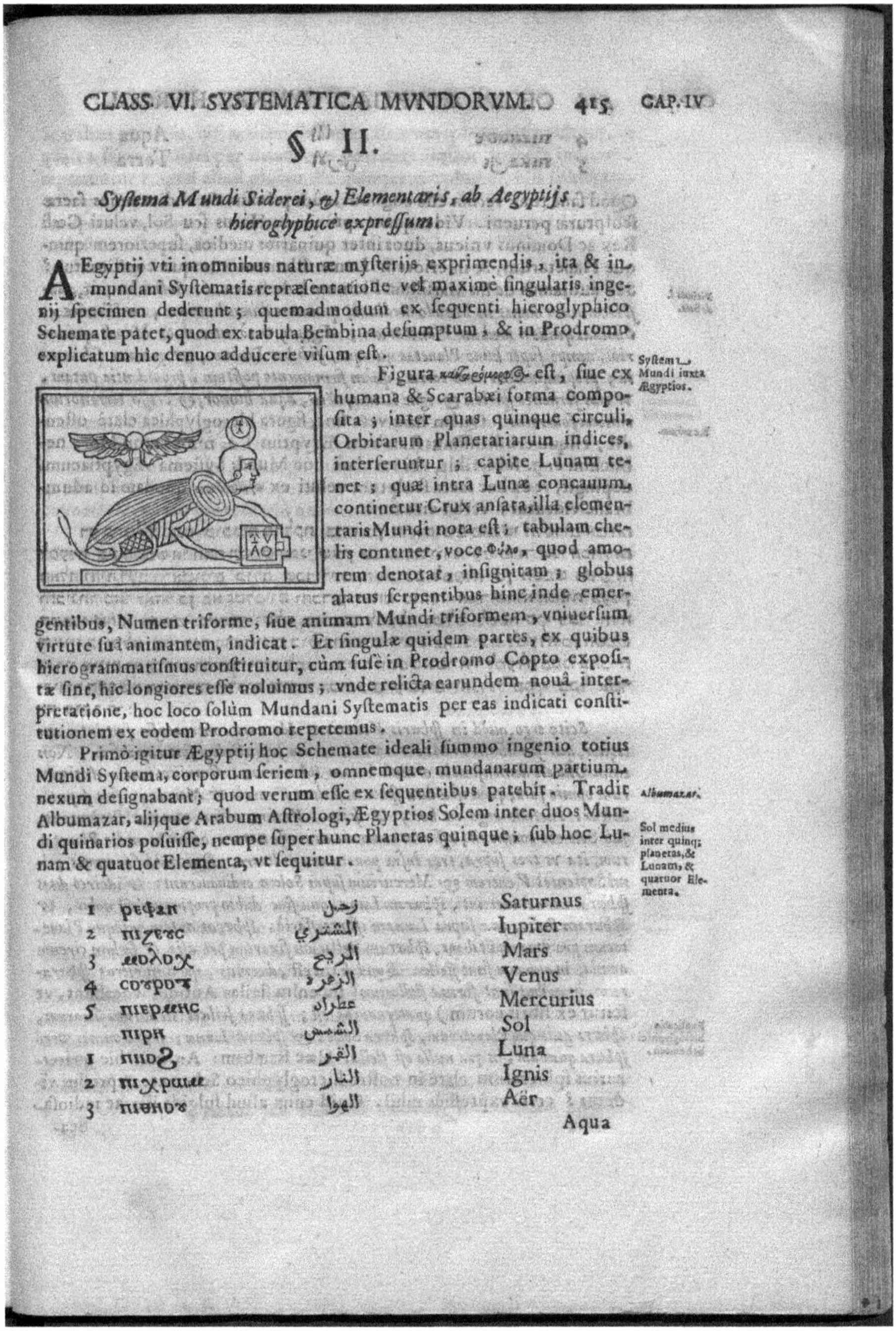

CLASS. VI. SYSTEMATICA MVNDORVM. 415 CAP. IV

§ II.

Syſtema Mundi Siderei, & Elementaris, ab Aegyptijs hieroglyphicè expreſſum.

AEgyptij vti in omnibus naturæ myſterijs exprimendis, ita & in mundani Syſtematis repræſentatione vel maximè ſingularis ingenij ſpecimen dederunt; quemadmodum ex ſequenti hieroglyphico Schemate patet, quod ex tabula Bembina deſumptum, & in Prodromo explicatum hic denuo adducere viſum eſt.

Figura καν[illegible] eſt, ſiue ex humana & Scarabæi forma compoſita; inter quas quinque circuli, Orbitarum Planetariarum indices, interſeruntur; capite Lunam tenet; quæ intra Lunæ concauum, continetur Crux anſata, illa elementaris Mundi nota eſt; tabulam chelis continet, voce Φιλο, quod amorem denotat, inſignitam; globus alatus ſerpentibus hinc inde emergentibus, Numen triforme, ſiue animam Mundi triformem, vniuerſum virtute ſuâ animantem, indicat. Et ſingulæ quidem partes, ex quibus hierogrammatiſmus conſtituitur, cùm fuſè in Prodromo Copto expoſitæ ſint, hic longiores eſſe noluimus; vnde relicta earundem nouâ interpretatione, hoc loco ſolùm Mundani Syſtematis per eas indicati conſtitutionem ex eodem Prodromo repetemus. Syſtema Mundi iuxta Ægyptios.

Primò igitur Ægyptij hoc Schemate ideali ſummo ingenio totius Mundi Syſtema, corporum ſeriem, omnemque mundanarum partium nexum deſignabant; quod verum eſſe ex ſequentibus patebit. Tradit Albumazar, alijque Arabum Aſtrologi, Ægyptios Solem inter duos Mundi quinarios poſuiſſe, nempe ſuper hunc Planetas quinque; ſub hoc Lunam & quatuor Elementa, vt ſequitur. Albumazar. Sol medius inter quinq; planetas, & Lunam, & quatuor Elementa.

1	ⲣⲉⲫⲁⲛ	زحل	Saturnus
2	ⲡⲓⲍⲉⲩⲥ	المشتري	Iupiter
3	ⲙⲟⲗⲟϫ	المريخ	Mars
4	ⲥⲟⲩⲣⲟⲧ	الزهرة	Venus
5	ⲡⲓⲉⲣⲙⲏⲥ	عطارد	Mercurius
	ⲡⲓⲣⲏ	الشمس	Sol
1	ⲡⲓⲟϩ	القمر	Luna
2	ⲡⲓϫⲣⲱⲙ	النار	Ignis
3	ⲡⲓⲑⲏⲟⲩ	الهوا	Aër
			Aqua

FIGURE 6.3 Kircher, Scarab-man with tablet, https://digital.library.cornell.edu/catalog/ss:550228.

were created as the same time as earth. He finds precedent for this notion in Genesis 1:11: "And God said, Let the earth bring forth grass, the herb yielding seed, and the fruit tree yielding fruit after his kind, whose seed is in itself, upon the earth; and it was so." Kircher argues that the notion of *panspermia* developed in ancient Egypt and is symbolized by the scarab beetle. In Rowland's translation of a treatise written during the early 1660s, the figure of a scarab is the figure of the sun-god "because of the similarity and analogy between the work of the beetle and the work of the sun" (Rowland 2005: 514). The scarab "gives life and fertility to his ball," explains Kircher: he rolls it towards the sun, gathering seed, while the sun-god, "orbiting the globe," gives the ball "life and fertility by the means of the same *panspermia rerum*, and fills it with every kind of thing" (Rowland 2005: 514).

It is a few short associative steps from the image of the life-giving sun-god to the idea of Jesus Christ, also the "only begotten," as a symbolic dung beetle, capable of giving life and of self-resurrection.

This odd, syncretic image of Jesus Christ as a dung beetle reveals the complex and unpredictable ways that certain insects become encrusted with layers of meaning. Kircher was aware of folk tradition in Germany, which connected the stag beetle, associated with Thor, the god of thunder and lightning, and Christ (Ratcliffe 2006: 91). The most famous artistic examples of this tradition come from Albrecht Dürer, who imbedded the lowly stag beetle in religious paintings, including *Madonna with a Multitude of Animals* (*c.* 1505) and *Adoration of the Magi* (1504). Dürer may or may not have been responding to the work of St. Ambrose, who, referring to the single mention of a beetle in the Bible, had compared the beetle to Christ. Kircher, in contrast, seizes on this association but reinterprets Christ's words in Psalm 22—"I am a worm, and no man"—by interpolating "scarab" for "worm" (Cambefort 1994: 13–14). Kircher's investment in this particular translation reflects his faith in an exegetical tradition to which, he believed, only the erudite have access. "Holy Scripture," he claimed, "is nothing other than an extended symbol of the most sublime matters and mysteries." Because "the hidden substance of God does not know how to enter profane and polluted ears by means of naked speech," Christ was forced to convey "eternal Wisdom" through parables and symbols: "Julian the Apostate, although impious, rightly said that 'Divine Nature loves to be covered and hidden away'" (Rowland 2002). In tracking down and exposing the network of associations through which Christ and the dung beetle could be allied symbolically, Kircher imagines himself as accessing a language of God available only to the most devoted and gifted of interpreters.

It is worth pointing out, in closing, that Kircher's defense of the symbolic method is diametrically opposed to the argument from design that dominated the discourse of seventeenth-century physico-theology. Whereas John Ray, Robert Boyle, and Isaac Newton, among others, saw the wisdom of God manifested in the physical world, including its lowliest creatures, Kircher embeds souls, symbols, and insects in networks of occult associations that require painstaking explication. Not surprisingly, then, within a decade of his death, Kircher's figurative rehabilitation of the dung beetle was mocked as a sign of an esoteric scholarship too invested in its own credulity-stretching interpretations. Yet his image of Christ as dung beetle still speaks powerfully to the complexity of insect symbology in the Renaissance. His appeal to intricate chains of intertextual relationships suggests the ways in which acts of symbolic interpretation depended on prior knowledges and modes of signification—on the Bible, and languages long dead—to which the learned, despite the widespread use of emblem books, had most access. Within this context, insects and their still-mysterious powers of reproduction were critical moral and affective metaphors for the exercise of existing social, religious, and political authority: as symbols of demonic possession, they were used to deny the female role in reproduction and to ground sociopolitical models in which women and poor were marginalized. At the same time, it would be a mistake to reduce the semiotic complexity of insects to stable guarantees of value. As I have argued in this essay, insect symbology helps define and shape Renaissance culture precisely to the extent that flies, moths, bees, and beetles resist one-size-fits-all interpretations. This is the lesson that Kircher's dung beetle (unwittingly) teaches, and the problem of emblematic meaning that later naturalists take pains to avoid.

CHAPTER SEVEN

Insects in Literature and Language

From Chaucer to Shakespeare

ANN M. MARTINEZ

Along the Rhineland, during the thirteenth and fourteenth centuries, a number of cathedrals showcased the sculpture of a beautiful woman or a handsome man with a dark side. The attractive and engaging appearance of both sculptures hid their true nature—their backside presented the depiction of putrefying flesh over-crawling with vermin, reptiles, and, of course, insects.[1] *Frau Welt*, or Lady World, and *Fürst der Welt*, Prince of the World, were popular medieval German allegories that showcased the temptation and corruption of earthly things. The lesson had a biblical origin: "Love not the world, neither the things that are in the world" (*nolite diligere mundum neque ea quae in mundo sunt*) (McDonald 2014: 628). The vehicle for the lesson was various creatures thought of as repugnant. Not surprisingly, "[t]he tempter figures conflate insects with the devil, deceit, and worldly desire, characterizing the threat to the wholeness of the body posed by sin" (Coutts 2006: 300). The figures are also a form of *memento mori*, reminders to onlookers of the death that awaits all humans and the decay that awaits the immoral and the evil. The medieval church stressed that the bodies of the righteous were safe from putrefaction, associating insects with punishment after death. But insects could also be proof of "god's immensity" as St. Augustine stated—as long, however, as it was the "right" kind of insect, like ants or bees (Johnston 2011: 40). Western European medieval literature praised or condemned, ignored or underscored insects in accordance with society's views, which were, in turn, heavily influenced by the church and cultural customs—a practice that continued well into the literature of the Renaissance.[2]

When the Black Death began making its way across mainland Europe in 1347, to say that everything changed is an understatement. By some estimates, a third of the population died thus shaking society to its core and altering all spheres of daily life, from the physical (food supplies, availability of work, educational practices, etc.) to the metaphysical (views on religion and death). In his *Epistolae de Rebus Familiaribus et Variae*, the Italian poet Francesco Petrarch lamented and questioned what he observed after the plague's onslaught. Writing from Parma in 1349, he asked, "Has what happened in these years ever been read about: empty houses, derelict cities, ruined estates, *fields strewn with cadavers*, a horrible and vast solitude encompassing the whole world?" (Aberth 2017: 60; emphasis added). It most certainly must have felt like the world was coming to an end. For his part, Italian author and humanist Giovanni Boccaccio, who described the Black Death in his famous work *The Decameron*, noted that the plague changed people's behavior toward each other. Gone were the common kindnesses toward the sick and ailing, as

> [m]any dropped dead in the open streets, both by day and by night, whilst a great many others, though dying in their own houses, drew their neighbors' attention to the fact more by *the smell of their rotting corpses* than by any other means. And what with these, and the others who were dying all over the city, bodies were here, there, and everywhere.
>
> (Aberth 2017: 66–7; emphasis added)

The proliferation of unburied cadavers did something more than mark the decimation of the population: it challenged the living's beliefs in good and evil by bringing insects to the forefront.

For survivors, the plague outbreaks led to the "gradual erosion of the unquestioned belief in the teachings of the church, which dominated the medieval world," and this new insight "was accompanied by a proliferation of insect imagery" (Coutts 2006: 300). The presence of insects in the decomposition process destroyed the hope for the preservation of the human body after death for those whose deeds in life were meritorious. Thus, the dark side of insect imagery began to widen. While in the past, sculptures of *Fürst der Welt* and *Frau Welt* were associated with evil, after the plague they were reminders of death in general as the sight of cadavers infested with insects became a common sight as Petrarch and Boccaccio noted. Michel Beheim's fifteenth-century song-poem on *Frau Welt*, titled *Ain beispel von ainem weib, was vorn schan und hinden schraglich* (A moral lesson about a woman, beautiful in front and frightful at the back), tells of a young knight enamored of Lady World. After consummating their relationship, Beheim emphasizes what the knight sees as he "catches sight of her ulcerous back. He sees maggots, worm-maws, pus-filled tissue, and dead, putrefying flesh and claims that he had never smelled anything to equal it" (McDonald 2014: 620). Such a sight might have been witnessed by many during the years of the Black Death. People had an understanding of the involvement of insects and "[m]ost educated men ... knew that insects were produced exclusively from corruption, spontaneously generated from rotting leaves, dead bodies, all sorts of putrefaction" (Manley 1961: 400), but the problem was in explaining their presence: either God made them intentionally, they created themselves from death, or they resulted from original sin. It is somewhat ironic to think that the high visibility of cadaver insects led to major shifts in cultural views, all stemming from the involvement of an almost invisible insect: the *Xenopsylla cheopis*, or Oriental rat flea, the primary vector of the bubonic plague. Although not identified until 1898 by Paul-Louis Simond, a flea, a tiny insect almost unnoticeable to the human eye, was the intermediary in the transmission of one form of the plague. It is hard to imagine that, had people in the Middle Ages known fleas were involved in the death and decimation around them, insects could have been judged any more negatively than they already were.

The inferior views toward insects in some cases went beyond the insect itself and extended to outside associations. Take, for example, the worm. Many insects, and other creatures, were classified as being *worms* as it was "a generic description for anything slithering and repulsive, up to and including a crocodile" (Johnston 2011: 38). The bestiary in MS Bodley 764 includes a classification of the different types of worms as acknowledged in the Middle Ages:

> The worm is an animal which is born of flesh or wood or any material thing without mating: but nonetheless worms are born from eggs. But they are either of the earth, of water, of air, of flesh, of leaves, of wood or of clothing.
>
> (Barber 1999: 199)

Regardless of their possible origins or material source, worms tainted their environs. In Geoffrey Chaucer's *The Parliament of Fowles*, a Middle English poem in which birds debate over love and the choice of partner before the figure of Nature, the "foules of ravine" (birds of prey) are the "hyest set" (highest ranking), while those who eat "worm or thing of which I telle no tale" (things of which I will not speak) are one level below (ll. 323, 324, 326). The birds, identified by what they eat are ranked accordingly. Similarly, in Chaucer's *The Legend of Good Women* when the God of Love sees Chaucer, who is accused of depicting women negatively, kneeling near a flower that represents women's faithfulness, the God chastises him.[3] He says:

> .... What dostow her
> So nygh myn oune floure, so boldely?
> Yt were better worthy, trewely,
> A worm to neghen ner my flour than thow.
>
> (ll. 315–18)

> (What are you doing here, so near my own flower, so boldly? It would be better, truly, for a worm to approach my flower than you.)

A worm would, of course, pose a direct threat to the flower bud but would do so for its own subsistence and not maliciously. The contrast between the man and the worm is meant to underscore the speaker's lowliness—it is hard to be lower than a worm, but he has managed to do so because of his actions. When the speaker asks for a reason behind the insult, the God of Love goes on to tell him, "For thow ... art therto nothing able" (For you are completely worthless) (l. 320) and "mysseyest" (slanders) (l. 323) those devoted to Love.

In the opening lines of *The Parliament of Fowles*, Chaucer describes a nature scene in great detail. He writes,

> A gardyn saw I ful of blosmy bowes
> Upon a river, in grene mede,
> There as swetnesse everemore inow is,
> With floures white, blewe, yelwe, and rede,
> And colde welle-stremes, nothyng dede,
> That swymmen ful of smale fishes lighte,
> With fynnes rede and skales sylver bryghte.
>
> On every now the bryddes herde I synge,
> With voys of aungel in here armonye;
> Some besyede hem here bryddes forth to brynge;
> The litel conyes to here pley gonne hye;
> And ferther al aboute I gan aspye
> The dredful ro, the buk, the hert and hynde,
> Squyrels, and bestes smale of gentul kynde.
>
> (ll. 183–96)

> (A garden saw I full of blossoming boughs above a river, in a green meadow, where sweetness is always plentiful, with flowers white, blue, yellow and

red, and with cold well-streams, nothing dead, which are full of fish, small and swift, with red fins and scales silver bright. On every bough I heard the birds sing with angelic voices in their harmony; some busied themselves to bring out their fledglings; and little rabbits to their play hurried by. And further all around I noticed the fearful roe, the buck, the hart, the hind, squirrels, and small beasts of a noble kind.)

Not a single insect is mentioned. Even though such a nature scene should include a diversity of insects, and certainly butterflies and bees, they are nowhere. It is interesting to note that Chaucer does not dwell much on insects in his literary works. He also, notably, does not discuss the plague much even though he was likely around five years old when it first struck London and as an adult lived through a number of outbreaks. One can speculate about the possible association between the plague and insects, and ponder as to why Chaucer may have wanted to erase both from his memory and only mention the occasional worm. When it came to insect exclusion Chaucer was not alone. There is a striking absence of insects in medieval literature, and when they do appear it is usually as a form of punishment, akin to the wasps and horseflies that sting and bite the Uncommitted sinners in the vestibule of Hell in Dante's *The Divine Comedy.*[4]

Medieval illuminated manuscripts were different, though. While insects are not often mentioned in the texts themselves, the marginal and intra-textual illustrations found in fourteenth- and fifteenth-century manuscripts often do depict insects and do so favorably. Open any extant breviary or book of hours, and one can find that butterflies were a favorite decorative addition alongside flowers, as can be seen, for example, in *The Hastings Hours*, where the marginalia around St. Margaret includes four butterflies, or in *The Breviary of Queen Isabella of Castile*, where, alongside a bagpipe-playing monkey one can spot a butterfly (Figures 7.1 and 7.2, respectively).[5] But butterflies had company every now and then from other winged insects. Bees and even flies also appear amidst the flowers. In one book of hours, a butterfly shares space with a bee and a bluish-tinted fly.

More strikingly, in *Hours of the Holy Spirit*, the only insect adorning the page is what looks to be a common black fly.[6] The insects that adorn the pages of the manuscripts are interspersed amidst beauty and thus associated with it rather than with death and pestilence. The lone fly might be an exception to this and have a transitory link to death since it appears in an image alongside the body of Christ as it is being entombed after the crucifixion. The design of manuscript illumination was very intentional and even pragmatic, with "[d]ecoration [meant as] a device to help a reader use a manuscript" (De Hamel 1986: 101), leaving clear identifiers for ease of navigation. In this sense, images of insects could have worked as sign posts. Even though many illuminations were fantastical in one way or another, bringing to mind the bagpipe-playing

FIGURE 7.1 *The Hastings Hours*, *c.* 1480, British Library, Add. MS 54782, f. 148 62v.

FIGURE 7.2 *The Breviary of Queen Isabella of Castile*, *c.* 1487, British Library, Add. MS 18851, f. 412v.

monkey, there was also a strong push toward realism and accurate depiction. In fact, a trend in Flemish books of hours was that of realistic borders that looked as if "flowers had actually fallen onto the pages" (De Hamel 1986: 169), as can be seen in *Book of Hours* MS 25698 and in *Hours of the Holy Spirit* (Figures 7.3 and 7.4, respectively). An attempt at realism might be the reason why these are two of the pages that depict insects other than butterflies.

If images of insects were more widely received, speaking of them was not so popular with medieval authors. While Chaucer avoids most insects, he does talk about bees. In *The House of Fame*, Chaucer, who is meant to be the narrator, describes a dream he has of visiting the palace of the goddess of Fame. There, as he looks around observing a group of authors of heroic tales, he hears something approaching. He writes,

> But while that y beheld thys syghte,
> I herde a noyse aprochen blyve,
> That ferde as been don in an hive
> Ayen her tyme of out-fleying;
> Ryght such a maner murmurynge,
> For al the world, hyt semed me.
>
> (ll. 1520–5)

> (But while I beheld this sight, I heard a noise approaching quickly that sounded as bees do in a hive just before flying out; just such a manner murmuring, for all the world, it seemed to me.)

The noise is not coming from bees, but from a "gret companye" (large group) of people from "sondry regiouns" (various regions) who were of "alleskynnes condiciouns" (all walks of life) there to beg boons from the goddess of Fame. Their association with bees gives the group a sense of activity, but more importantly of anticipation and hope as one might gather from bees eager to leave the hive and start the day's work. The analogy is not a damning one, as it might have been had the comparison drawn a connection to the buzzing of flies. Beekeeping during the Middle Ages was such a flourishing practice, with both honey and wax highly prized, that laws and regulations were put in place to encourage and protect it: Charlemagne decreed that all manors keep bees, and during King Alfred's reign the "highest fines" were imposed on the "'stealer of bees'" (Kritsky 2017: 253). Such an appreciation did not wane, as can be seen from Thomas Moffet's assertion in the *Theatrum insectorum* (*c.* 1589): "Who would not acknowledge for truth ... that the Bee hath in it a particle of divine understanding, and heavenly wisdom? Who I say will deny them to have fantasie, memory and some kinde of reason" (qtd. in Johnston 2011: 38) (see Figure 7.5).

FIGURE 7.3 *Book of Hours*, Netherlands, *c.* 1492, British Library, Add. MS 25698, f. 3.

FIGURE 7.4 From the *Hours of the Holy Spirit*, Netherlands, *c.* 1500, British Library, Add. MS 35313, f. 31.

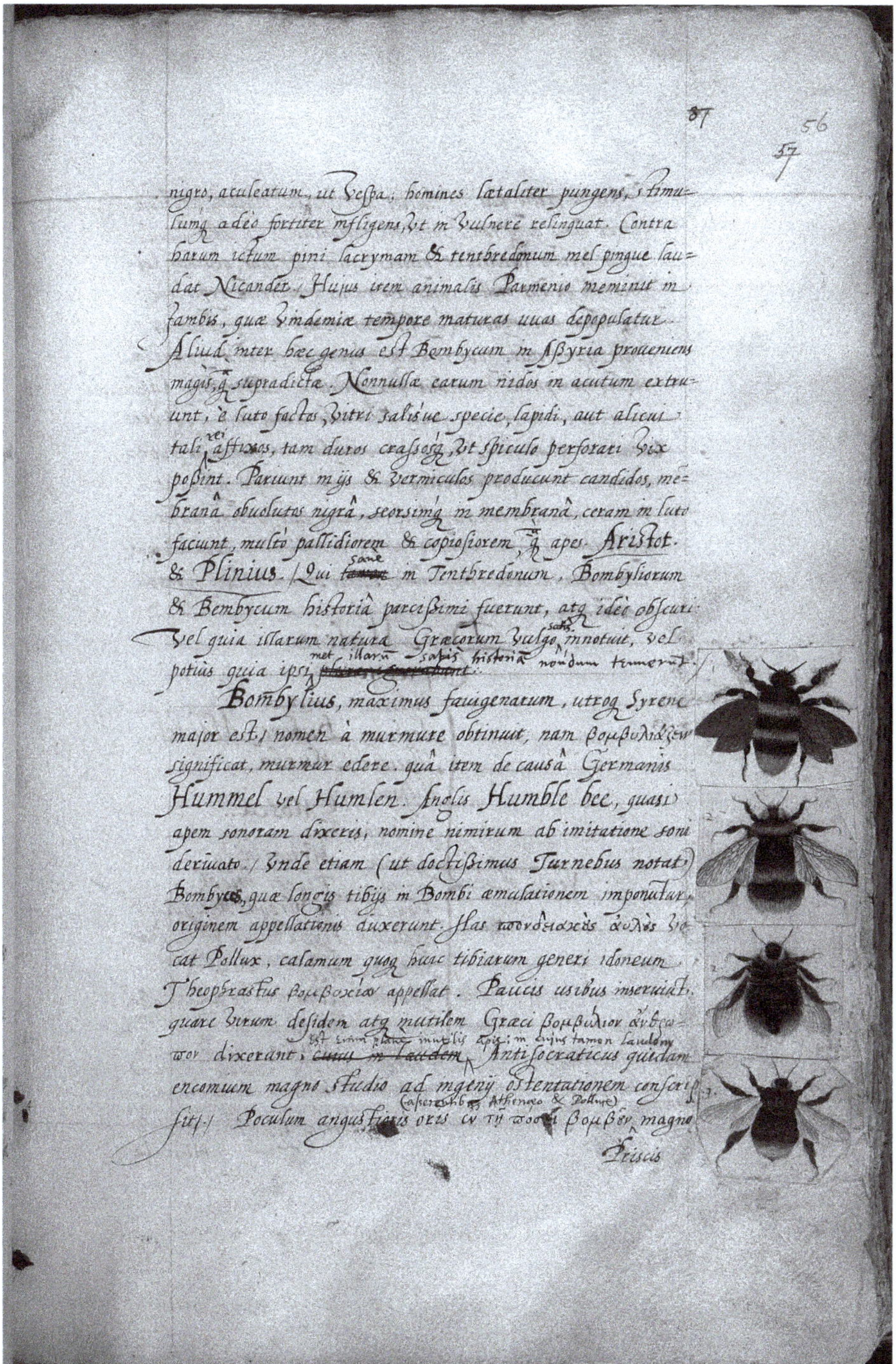

FIGURE 7.5 *Insectorum theatrum*, British Library, Add. MS Sloane 741 (4014), f. 56.

Dante, for one, would have agreed with Moffet, having placed bees in Heaven in *The Divine Comedy*. In the bestiary found in MS Bodley 764, bees are even praised for their gruesome although incorrect birth simply because it is not sexual:

> They are known, by observation, to be born from corpses of oxen. In order to obtain bees, the flesh of dead calves is beaten; from its putrefying blood worms emerge, which later become bees. So the insects that come from cattle are called bees; those that come from horses are hornets; mules produce drones and donkeys bring forth wasps.
>
> (Barber 1999: 178)

Such an origin story speaks directly to the power of bees, and a few other winged insects, since their value and appreciation did not diminish even when thought to come from death.

Things are no different in the fourteenth-century Middle English poem *Sir Gawain and the Green Knight*. Although the poem focuses heavily on scenes of nature and includes one of the most famous hunting scenes of all medieval Romance, attention in nature is given primarily to vegetation, birds, the animals of the hunt, vermin, and even some monstrous creatures. There is, however, a single reference to an insect: the butterfly. When King Arthur and the Knights of the Round Table are gathered together in Camelot celebrating the Christmas holidays, an extremely large green man rides in on a green horse and halts festivities. He offers up a challenge, known as a "beheading game," to those present in exchange for a beautiful new ax. While the Green Knight might appear threatening precisely because of his challenge, his attire indicates the opposite. Rather than battle armor, he is wearing an elegant outfit richly adorned with meticulous detail:

> That were to tor for to telle of tryfles the halve
> That were enbrauded abof, wyth bryddes and flyyes,[7]
> With gay gaudi of grene, the golde ay inmyddes.
>
> (ll. 165–7)

> (It would be too difficult to tell of half the details that were embroidered upon, the birds and butterflies, with green ornamentation centered in gold.)

The single reference to an insect that appears in *Sir Gawain and the Green Knight* does not even refer to an actual insect, but instead refers to the visual representation of one on a character's clothing. A later description of Sir Gawain's armor shows that he, too, has decorative elements from nature. However, they are not insects. In his case, the *vrysoun* on his helmet is embroidered with "papjayz" (popinjays) (l. 611). Of note is the fact that in this poem King Arthur's court is "indoor-focused," heavily contrasted with the Green Knight's outdoor-focused estate where he "exhibits a concern for his land and its creatures that shows an early English example of environmental awareness" (Martinez 2016: 114). The popinjays on Gawain's clothing are a "stylized nature, an idealized beauty constructed by humans, [and] based on a human romanticized view of nature"—a nature devoid of insects (Martinez 2016: 123). For the Green Knight, whom some scholars have argued is a representation of the Green Man, and who has a much more direct interaction with nature, the inclusion of an insect in his attire depicting nature seems almost obligatory. These butterflies, like the embroidered birds they accompany, are meant to be beautiful and elicit a sense of sumptuousness, elegance, and nobility in the clothing and in the person wearing the garb. Here the butterfly is an insect appreciated for its beauty and not for its material contribution to humans (e.g., honey, wax) or for its work ethic (e.g., bees, ants). Beauty for the sake of beauty can be found in an insect.

Based on the extant literature from the period, for medieval authors insects could be weaponized, used by a higher power as a form of punishment against humans in life and after death; they could be instruments, teaching moral lessons and human values to those willing to learn and correct their path; and once in a while they could be beautiful. For Renaissance authors not much changed, except for the fact that they were much more willing to talk about insects. Shakespeare was one of these authors.[8] In the mid-nineteenth century, Irish naturalist Robert Patterson took inventory of Shakespeare's works, transcribing all the passages containing references to nature, "and found that they occupied one hundred closely written pages of letter paper." Of these, nine pages were related to insects (Patterson 1838: 11).

Like his medieval predecessors, Shakespeare primarily included references to select flying insects and to worms. However, his use of insect analogies fluctuates between positive associations and damning ones without a clear pattern as was found before. In *Two Gentlemen of Verona*, a play many scholars consider to be Shakespeare's first, wasps and bees are placed in opposition to each other early on. Julia and her waiting-woman, Lucetta, are having a back-and-forth regarding a love note Proteus has sent Julia. In an attempt to disregard her feelings toward him, Julia rips up the letter and scatters the pieces on the ground. Once Lucetta leaves the room, Julia laments her own actions, immediately picking up pieces of the torn letter:

> O hateful hands to tear such loving words;
> Injurious wasps to feed on such sweet honey
> And kill the bees that yield it with your stings!
>
> (1.2.106–8)

As instruments of destruction, her hands become wasps and the letter itself the bees as it carries honeyed words of love to her. Two to three years later, when Shakespeare wrote his most violent and gruesome play, *Titus Andronicus*, bees were no longer analogized as messengers of love but instead instruments of war. When Lucius joins the Goth army to exact revenge on Tamora for destroying his family, one of his new allies talks of his enthusiasm for what is to come:

> Brave slip sprung from the great Andronicus,
> Whose name was once our terror, now our comfort,
> Whose high exploits and honorable deeds
> Ingrateful Rome requites with foul contempt,
> Be bold in us. We'll follow where thou lead'st,
> Like stinging bees in hottest summer's day
> Led by their master to the flowered fields,
> And be avenged on cursèd Tamora.
>
> (5.1.9–16)

The Goth soldier pledges the Goth army to Lucius by comparing themselves to bees. While fighting by stinging on the hottest day of the year does add a degree of hostility to the analogy, it appears more poetic than threatening when he mentions fighting in a field of flowers. The threat also is diminished by our knowledge that once a bee stings it dies. Whether or not this fact was known to a Renaissance audience is unclear, since the Goth soldier's words could also be taken as his willingness to sacrifice himself and the rest of the army in order to defeat Tamora. However, there were a number of misconceptions regarding bees during the Middle Ages and the Renaissance. For one, people believed that bees were led by a "king" rather than a queen bee in part because there was a tendency to politicize the structure of the beehive.

The highly structured and hierarchical organization of bees opened the door for them to be anthropomorphized. As Christopher Hollingsworth explains,

> everything had its proper place in the Great Chain of Being. Because we aspire to an image of its politics, we saw the natural beehive as kin. Rhetoric not biology, analogy rather than blood, tied us to this living allegory, which signified visually in a language that all could understand.
>
> (Hollingsworth 2001: 152)

It was something people could relate to and thus extrapolate their own socio-political understanding unto a flying insect. In the MS Bodley 764 bestiary, the entry for the bee mentions that "[t]hey choose a king, create their own people, and although subject to a king are nonetheless free," in a show of wishful analogizing by the scribe (Barber 1999: 178). For his part, the bee "king" "has a sting, but does not use it in revenge; for there are laws of nature, not written down in letters, but innate in custom, that those who have unlimited power should be all the milder in their punishment" (Barber 1999: 179). The Goth soldier's words in *Titus Andronicus* likely reflected the day's general understanding about bees since the bestiary also mentions that bees "have an army" and "they fight battles" (Barber 1999: 177).

Shakespeare, an expert in the art of insults, often used insects as a vehicle for criticism. Gnats, worms, fleas, and wasps were common in his language arsenal for such a purpose, but even more so was the caterpillar. Even though he offered up politically pointed commentary, Shakespeare was an author who succeeded at avoiding serious conflict with the monarchy during his lifetime.[9] His success likely is the product of his masterful use of language and analogy. In *Richard II*, when a young Bolingbroke, the future King Henry IV, is recruiting royal family members for his cause against King Richard, he uses logical associations from nature to elicit a response. He says to York, his uncle:

> But we must win your grace to go with us
> To Bristol Castle, which they say is held
> By Bushy, Bagot, and their complices,
> The caterpillars of the commonwealth,
> Which I have sworn to weed and pluck away.
>
> (2.3.161–6)

Bushy and Bagot are two of the three supporters and advisors of King Richard who are accused of misleading him for their own benefit. Many blame them and their ill advice in part for the financial struggles and mismanagement the English people suffer through. They are rapaciously eating through the kingdom, just as a caterpillar would do. Bolingbroke is not alone in his use of the insect analogy. Toward the end of the play, while at their work one servant protests against the instructions the gardener has given him questioning the purpose of ordering the garden when the whole nation is in ruins. He says,

> When our sea-walled garden, the whole land,
> Is full of weeds, her fairest flowers choked up,
> Her fruit trees all unpruned, her hedges ruined,
> Her knots disordered, and her wholesome herbs
> Swarming with caterpillars?
>
> (3.4.43–7)

Known for being voracious feeders and at times agricultural threats, caterpillars were thought of as pests rather than future butterflies by a Renaissance audience. Even silkworms, which were valued for their production of silk, were regularly critiqued in apiculture and sericulture manuals for their morally dubious lifestyle since they simply lounged and fed all day, unlike the hardworking bees (Olbricht 2006: 223).

Bees, it seems, are always the standard against which to measure all others, including people. In Shakespeare's *A Midsummer Night's Dream* bees give insight into a person's behavior. This is a play that has invariably been associated with nature. For one, the main protagonists run away into the forest, but the presence of faeries also expands on the connection. The performance history of the play is riddled with visuals of nature, and often the King and Queen of the Faeries, Oberon and Titania, are dressed in nature-centric garb covered in flowers and vines. An early scene in the play shows Titania attempting to reason with Oberon since their quarrel has caused countless environmental mishaps—their discord has caused excessive flooding which has in turn led to failed crops and animal deaths (2.1.81–117). Her concern for the natural world around her dissipates once Titania is under the effects of a love potion and enamored of Bottom. Titania instructs her faery attendants to serve Bottom, a man recently transfigured into an ass. She tells them to feed him with various fruits, and

> The honey bags steal from the humble-bees,
> And for night-tapers crop their waxen thighs,
> And light them at the fiery glowworms' eyes
> To have my love to bed and to arise;
> And pluck the wings from painted butterflies
> To fan the moonbeams from his sleeping eyes.
>
> (3.1.150–5)

Shakespeare did not select insects that are considered pests, but instead chooses insects valued for their work and their beauty, the bee and the butterfly. Titania's willingness to steal from and dismember these two insects in the name of love can be read as a critique of the dangers of excessive emotion.

Titania's behavior toward winged insects is in stark contrast to that of the eponymous tragic hero in *Titus Andronicus*. Titus, a highly decorated Roman general in Shakespeare's first tragedy, finds himself at an emotional and psychological low point in his life: two of his sons have been wrongfully executed; a third son has self-banished to stay alive; Titus has lost his hand in a moment of sacrifice meant to save his family; his son-in-law has been murdered; and his daughter has been raped and mutilated. After this series of events, Titus' brother Marcus kills a fly during dinner. Titus immediately chastises his brother for his actions, crying "Out on thee, murderer! Thou kill'st my heart" (3.2.54) in a show of humanity toward a lowly insect he declares "innocent" (3.2.56).[10] When Marcus protests that he has "but killed a fly" (3.2.59), Titus responds:

"But"? How if that fly had a father and mother?
How would he hang his slender gilded wings
And buzz lamenting doings in the air!
Poor harmless fly,
That, with his pretty buzzing melody,
Came here to make us merry, and thou hast killed him.

(3.2.60–5)

Titus' description of the fly might remind the reader of a bee, often seen in a much more favorable light than that of a common fly. But Shakespeare's use of the fly in this scene underscores the emotional plight Titus is going through. Titus, who at the start of the play killed one of his own sons for disobeying him, has had his views on life and death altered by the loss and suffering he recently endured to where he now values even the life of a fly. For Titus and the audience seeing the play, the humanized fly is a stand in for the bonds of family.

Shakespeare does not abandon use of this winged insect in subsequent plays, for the "ill-favored fly" (3.2.66) as Marcus described it, serves a solid purpose to underscore human emotion and vulnerability. For Romeo, in *Romeo and Juliet*, a fly becomes an enviable stand-in for his love for Juliet. After Romeo has killed Tybalt, Juliet's cousin, during a fight, he is banished from Verona. While Friar Laurence tries to console him by reminding Romeo that he has not been sentenced to death for his actions, the young man insists that exile is worse than death because he will not be able to see his beloved. His torture will be knowing that all other creatures, including "every unworthy thing" (3.3.31) will be able to see her while he cannot. He goes on to say,

.... More validity,
More honorable state, more courtship, lives
In carrion flies than Romeo. They may seize
On the white wonder of dear Juliet's hand,
And steal immortal blessing from her lips,
Who even in pure and vestal modesty
Still blush, as thinking their own kisses sin.
This may flies do, when I from this must fly—
And sayest thou yet that exile is not death?
But Romeo may not; he is banishèd.
Flies may do this, but I from this must fly;
They are free men, but I am banished.

(3.3.33–44)

The mournful Romeo seems to believe that Juliet's beauty is such that it will attract carrion flies, or blowflies, even though these insects are primarily attracted to decomposing cadavers or rotting foods. He is even jealous of the blowflies who will have the opportunity to touch and kiss her lips. As he considers the proximity of the flies to his beloved's skin, he gets more and more restless at the thought that he will be away from her, imprisoned in a sense, while the flies have now become "free men" in his mind—able to approach her at any time. While Titus Andronicus gives the fly a family and the kind personality of one intentionally sharing its buzzing sound with humans, Romeo elevates the blowfly to the status of a personally and politically free man who does not live under totalitarian rule. While his use of carrion flies in his lament is meant to showcase his breaking heart, it is difficult to surmise what Juliet might have thought about such a notion, but she likely would not have considered it romantic.

Shakespeare's use of insects in courtship is by no means the most famous one. That title is reserved for John Donne's poem "The Flea."[11] Considered an erotic metaphysical poem, it is intended to convince the beloved that she and the speaker should consummate their relationship. Donne writes,

Mark but this flea, and mark in this,
How little that which thou deny'st me is;
It sucked me first, and now sucks thee,
And in this flea our two bloods mingled be;
Thou know'st that this cannot be said
A sin, nor shame, nor loss of maidenhead,
Yet this enjoys before it woo,
And pampered swells with one blood made of two,
And this, alas, is more than we would do.

(ll. 1–9)

In the opening stanza, the speaker argues to his beloved that in biting both of them, the flea has mixed their blood already which is what she is denying him in refusing his sexual advances. During this period, the general belief was that sexual intercourse led to the mixing of the participants' blood. In the following stanza, when the beloved attempts to squash the flea he tries to stop her hand by saying,

Oh stay, three lives in one flea spare,
Where we almost, nay more than married are;
This flea is you and I, and this
Our marriage bed, and marriage temple is;
Though parents grudge, and you, we're met
And cloistered in these living walls of jet.

Though use make you apt to kill me,
Let not to that, self murder added be,
And sacrilege, three sins in killing three.

The union of their blood in the flea, he claims, has led to their marital union both in the physical sense but also the spiritual. The flea has gone from representing the physical union itself to becoming a stand-in for the religious aspect of marriage: the flea is a "temple" to marriage. Additionally, Donne's choice of the word "cloistered" brings to mind religious orders and their dwellings. The author also elevates the argument further by accusing her of causing three deaths if she were to kill the flea. But his argument is unsuccessful, as he adds,

Cruel and sudden, hast thou since
Purpled thy nail in blood of innocence?
Wherein could this flea guilty be,
Except in that drop which it sucked from thee?
Yet thou triumph'st, and say'st that thou
Find'st not thy self, nor me the weaker now;
'Tis true, then learn how false, fears be;
Just so much honor, when thou yield'st to me,
Will waste, as this flea's death took life from thee.

As Titus before him proclaimed the innocence of the fly his brother Marcus killed, the speaker in the poem highlights the guiltlessness of the flea that his beloved has crushed with her nail. But when she assures him she does not feel affected by the death (since part of her now lived inside the flea), then he attempts once more to revive his case by reminding her that fears can be unfounded and that she will not feel any weaker or any lesser after she accepts his advances.

Donne's poem is famous primarily because of his unconventional metaphor. He is using the imagery of a flea to attempt to seduce someone. The flea might have a double purpose, though, since the poem also falls under the category of *carpe diem*, seize the day. The undercurrent in such poetry is that life should be lived in the moment since all must eventually die; in a sense, it is reminiscent of the *memento mori* prominent during the Middle Ages with its inclusion of insect imagery and subsequent association with death. Even when speaking of seduction, it seems the presence of an insect cannot but hearken toward death.

Insects are malleable, reflecting a society's views at any given time. Their association with punishment and death, prevalent in the Middle Ages and due in part to religious views and the onset of the plague, has never completely gone away, and they continue to "evoke frightening, dark, weird, and creepy sensations" (Berenbaum 2006: 6). But their presence also altered society's views regarding larger metaphysical questions on life, sin, good, and evil. While Renaissance authors, like Shakespeare, continued using insects in a similar way to that of their predecessors, later literary periods begin relying on the insect much more. By the time the Romantics appear, poets like Shelley are including references to twenty-two different insects in their work (Jeffrey 1976: 103).

CHAPTER EIGHT

Insects in Art

Art as Culture, Art as Science

CYNTHIA M. PYLE

> You who think to reveal the figure of man in words [alone]
> with his limbs arranged in all their different attitudes,
> banish the idea from you, for the more minute
> your description the more you will confuse the mind
> of the reader and the more you will lead him away
> from the knowledge of the thing described.
> It is necessary therefore for you to *represent and describe.*
>
> —Leonardo da Vinci, *Dell'Anatomia* (emphasis added)

In keeping with the recent development of concepts like "cultural entomology" (cf. Morris 2004: 181–215), this chapter will consider the various aspects of the role of insects in relation to humankind, specifically their relation

It is a pleasure to acknowledge the help of several people with updates on information, suggestions, bibliography, and advice on obtaining images: Dr. Gail Aronow; Dr. Holger Funk; Fabric Conservator Shelley Greenspan; Dr. Peter Jakab of the Smithsonian Institution; Professor Ingrid Rowland of the University of Notre Dame, Rome; Professor Jeffrey Chipps Smith of the University of Texas, Austin; Dr. Anna Spiro; as well as Bernadette Myers and Ana-Sofia Meneses, who assisted in the gathering of the images, and the editor of the series, Professor Gene Kritsky, and of this volume, Professor Sherilyn Smith of Le Moyne College, Syracuse, New York. The librarians at the British Library, Columbia University Libraries, the Frick Art Reference Library as well as the staff and curators of the Center for Advanced Study in the Visual Arts and the Library of Congress, both in Washington, DC, were helpful and accommodating.

to those quintessential artifacts of humanity, grouped together under the term "art." However, we will not be concerned here only with still-life paintings and other art works which include insects (and which burgeon in the seventeenth century, after the period, 1300–1600, covered here), but rather primarily with the uses of art as a tool for study of and to report findings about insects (among many other things: Pyle 2000). Any biologist knows how important this tool is. As the beautiful recent exhibition of the neurological drawings of Santiago Ramón y Cajal demonstrated, the very exercise of attempting to reproduce what one sees through the microscope is an active form of study (Ramón y Cajal 2017; cf. Hooke 1665). To fully understand a biological—or any—structure, one must draw it and describe it in words, as Leonardo da Vinci (1452–1519) understood. Drawing, then, can be considered a form of study, even of scientific analysis and description, as can be seen in fifteenth-century works like Leonardo's and Albrecht Dürer's (1471–1528) own highly accurate (and individually observed and produced, *pace* Ogilvie 2006 and Neri 2011: xix) drawings from nature, of everything from fossils to grasses to beetles.

Dyes and pigments are particularly relevant to our interest here, for the prized red pigments used in the decoration of precious manuscripts and later printed books, in the dyeing of fabrics, and also in paints include pigments derived from certain insects found in prehistoric times in the Near East, for example the scale insects (*Porphyrophora hamelii*) in Ararat, and, by the sixteenth century, in the New World cochineal (*Dactylopius coccus*) (Phipps 2010)—both in the order Hemiptera.

These and other brilliant pigments were chosen to depict the insects and other animals used to decorate (rather than illustrate) books of hours, breviaries and missals, with beautifully drawn and colored plants, insects and other "lowly creatures" in the margins (Figure 8.1).

The insects and animals chosen often included especially those with aposematic, or warning, coloration and patterns (see, e.g., Hutchinson 1974). Such colors evolved in certain poisonous butterflies, beetles, arachnids, and even reptiles and marine animals to ward off predators, which in turn have evolved to avoid them. Some species ingest toxins from plants or insects and incorporate them into their own physiology; natural selection then demonstrates the advantage of such incorporation, and, conversely, of its recognition by predators.

Most books that included animal depictions during the Middle Ages, and even well into the Renaissance (which can be considered to date from the fourteenth through the sixteenth centuries), were precisely such books of hours, missals and breviaries, or prayer books, embellished in their margins with sometimes very clear and identifiable depictions of animals, as reminders of the creation, or as allegorical memoranda. But, while these model book depictions were sometimes extremely accurate, and could in certain cases be seen (at least in the model books from which they were probably most often taken) to be careful

FIGURE 8.1 Breviario de Isabel la Católica [Manuscrito], fifteenth century (after 1495?), Biblioteca Nacional de España, Madrid MS VITR/18/8, fol. 37v, detail.

and specific studies of anatomy and behavior, their intention was not to serve as scientific studies or as illustrations to a text, but rather as useful repertories of figures for artists in their various compositions, sometimes allegorical.

Exceptionally, in medieval western Europe, the Holy Roman Emperor Frederick II of Sicily (1194–1250) wrote an empirical study of falcons and falconry, *De arte venandi cum avibus*, for his son Manfred. Significantly, this thirteenth-century manuscript (Palatinus latinus 1071, *c.* 1258–1266) was copiously *illustrated* (rather than decorated) with clearly identifiable birds, and annotated. The birds—both intended prey and falcons for the hunt—are indeed well described and clearly intended to illustrate the text, although this was not the norm in the West at the time. It is even possible that Frederick himself was the original observing artist, though he could not have painted the birds in this manuscript (Haskins 1921: 335, 339). Some birds, like the *hirundo rustica*, or swallow, in the upper right-hand corner of fol. 9, are seen in the act of capturing prey, in this case insects in flight (see: http://digi.ub.uni-heidelberg.de/diglit/bav_pal_lat_1071/0021/image). Unusually for western Europe (though not for Byzantium), this was an illustrated instruction book, akin to medical works but about hunting, and it did not emphasize moralization or metaphor, unlike the medieval *Physiologus* or its descendants, the popular bestiaries with their religious, moral, and ethical analogies and import.

Ancient and medieval encyclopedias, like those written up through the thirteenth century by Pliny the Elder (23–79 CE), Isidore of Seville (*c.* 560–636 CE), Vincent of Beauvais (*c.* 1184/94– *c.* 1264), Thomas of Cantimpré (or de Brabant, 1200/02–1263/72), and Albertus Magnus (*c.* 1200–1280), while differing in style and character among themselves, had in common with the *Physiologus* and bestiaries the intent to propose animals as moral exempla of behavior. Thus, Thomas of Cantimpré wrote a treatise on the bee (Cantimpré 1997), using it as a metaphor of monastic life, the "king" bee (as the queen was thought to be until Jan Swammerdam's dissection of the central figure in the seventeenth century [Cobb 2000]) representing the Abbot, and the drones the monks, silent and obedient. This metaphor has also been used for the state, and bees more generally have been seen as industrious and productive (of honey), from biblical times forward.

During the High Middle Ages, the term "*Vermes*" (best translated as "Lowly Creatures") was used to encompass many different animal forms, from actual worms (*vermes*) to insect larvae, such as caterpillars, to adult insects, and even frogs, toads, lizards, snails, and leeches. All these appear in the next to last book of a work, *De animantium naturis (On the Nature of Animals)*, by the fifteenth-century northern Italian humanist scholar Pier Candido Decembrio of Lombardy (1399–1477), who based his text on Thomas de Cantimpré's *De natura rerum*, which he had discovered as an anonymous text among the manuscript books of a nobleman's library in Naples in the late 1450s (Pyle 1984, 1996a, 2005).

Cantimpré's encyclopedia had also been used by Albertus Magnus in his *De animalibus*, and formed the principal source for Konrad von Megenberg's vernacular *Buch der Natur* of the early fourteenth century, which exists in a fair number of manuscripts, and was first printed by Johann Bämler of Augsburg on October 30, 1475. In this incunabulum, insects and other "lowly creatures" are recognizably represented, though in woodcut, and in a rather crude fashion (Figure 8.2). One can identify bees by their hives, silkworms by their round box, and locusts by their shape and wings. Such encyclopedic works differed from the books of hours, breviaries and missals, which were, as we have seen, decorated (rather than illustrated) with beautifully drawn and colored plants, insects and other creatures in the margins.

Medieval bestiaries are again different from either the breviaries and missals mentioned above, or these encyclopedic works; they especially differ from the natural history books one will begin to see in Germany and Italy in the late fifteenth century. These are books of moralized tales based on animal allegories from the Alexandrian Greek *Physiologus* of about the second century CE. They are usually illustrated (in the sense that the moral tales are reinforced by the drawings or woodcuts) with often fantastic creatures, modeled on medieval sculptures or drawings, whose humanoid physiognomies recur throughout

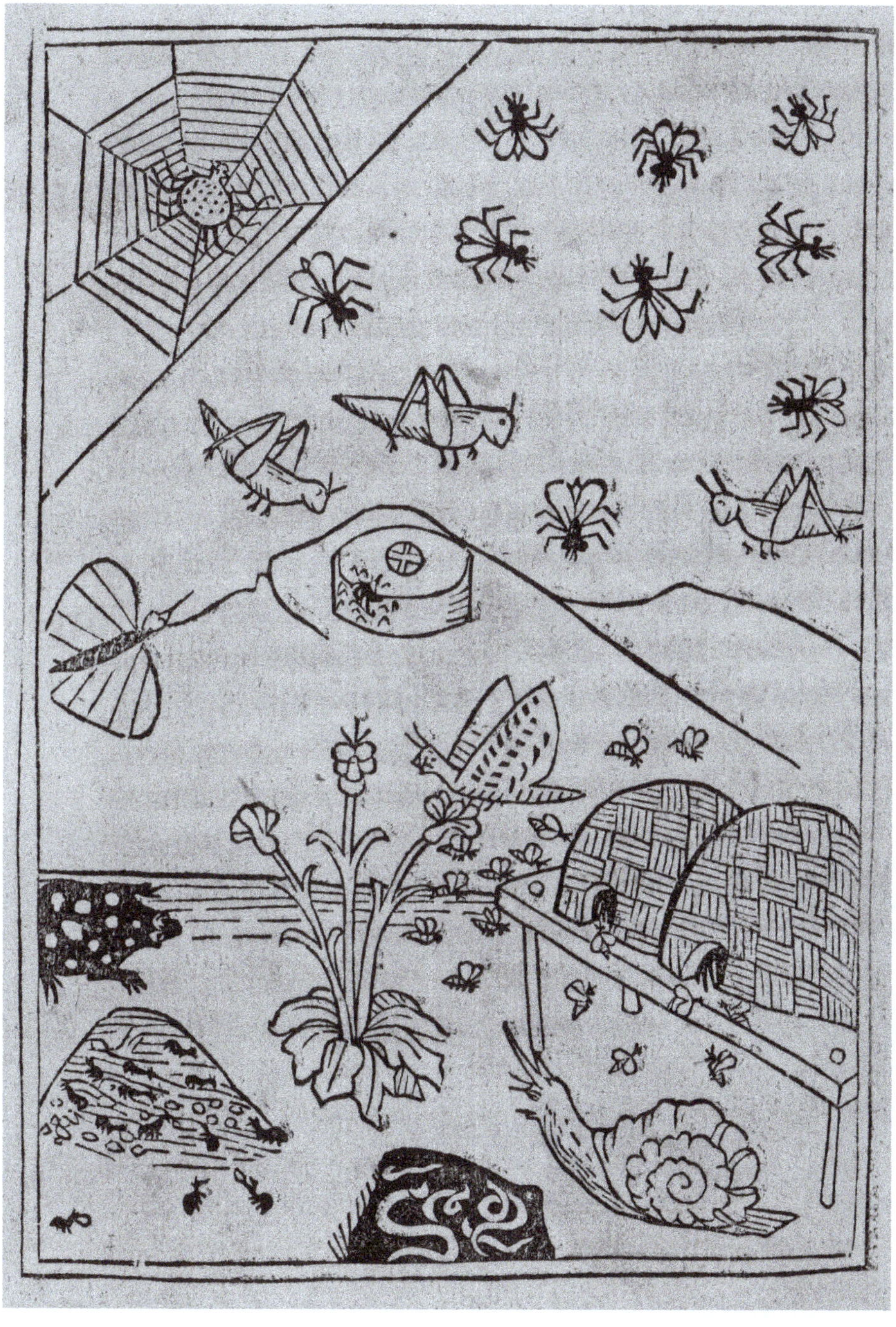

FIGURE 8.2 Konrad von Megenberg, *Das Puch der Natur*, Augsburg: Johann Bämler, 1475. (Huntington Library.)

the early modern natural history and artistic tradition. Their varying numbers of chapters, usually beginning with the lion, emphasize, first, the etymology of the name of a creature, then its description and habitat; the texts almost invariably conclude with a moral interpretation of the animal's appearance and behavior.

SOMETHING DIFFERENT EMERGES IN THE RENAISSANCE

In fourteenth- and fifteenth-century drawing, painting, graphic arts, and sculpture, particularly in Italy, attention to metaphor began to give way to attention to study of the real world, as the Italian philologists' and historians' approach shifted to the establishment of real texts and their real historical subject-matter. Now the emphasis was on learning about those realities. Likewise, the emphasis in some aspects of art was on learning and depicting the reality of the world about us. This approach was imitated from that found in certain newly recovered, and even previously known, works of antiquity. While the most qualified Italian humanists did not slavishly imitate their models, they did imitate these models' empirical and critical *approach* to the study of the world, in all its manifestations (Pyle 2006).

In various areas of learning, then, the fourteenth and fifteenth centuries in Italy began to demand more of their scholars, beginning in the time of the poet and humanist scholar Francesco Petrarca (or Petrarch, 1304–1374), and to a lesser extent even before. They began to require more accurate historical knowledge and historical understanding of the context surrounding the continually discovered manuscripts, ancient texts, and artifacts from ancient societies, such as those partially uncovered since antiquity in Rome and other ancient cities, which the great architect and artist Leon Battista Alberti (1404–1472) actually measured, mapped, and described. No longer was it felt necessary to justify all learning by its links to religious learning. That is not to say that religion disappeared from the social and cultural context—far from it—merely that the new emphasis in scholarship and the arts was on an empirical recognition of the actual world as it appears to us, and as we interact with its elements and phenomena here and now.

Philological techniques were developed to begin to establish ancient texts in keeping with the increasing knowledge of classical authors and their times. Scholars like Petrarch, Lorenzo Valla (1407–1457) and Angelo Poliziano (1454–1494) began this work, which would be carried on by later scholars in following centuries. At the same time, the realism seen in the decoration of manuscript prayer books and breviaries also came to dominate painting and sculpture, based on the same empirical requirements of the humanist scholars and the new taste they were creating in European culture. Angelo Poliziano taught the young Michelangelo Buonarotti (1475–1564) in the Medici household, instilling—or finding and cultivating—in him this same taste for empirical reality. Albrecht Dürer in the north too was in touch with the humanist scholars of his age, like Desiderius Erasmus (1466/69–1536, whose portrait he drew several times and then engraved), and he partook of this increasingly intense attempt to understand the world around him through his art, or *techne*.

At the same time, other artists were beginning to revel in and capture the reality of the world around them, both for its own sake and metaphorically, including as a *memento mori*. For example, a single, accurately described horsefly can be found on the windowsill of the *Portrait of a Carthusian* by Petrus Christus of 1446 (Metropolitan Museum of Art: https://www.metmuseum.org/art/collection/search/435896), and an equally accurate housefly is poised on the coif of a woman of the Hofer family in a Swabian artist's portrait of about 1470 (https://www.nationalgallery.org.uk/paintings/swabian-portrait-of-a-woman-of-the-hofer-family; Chastel 1984: 22–3).

In 1493, Hartmann Schedel (1440–1514) created and underwrote the famous printed *Nuremberg Chronicle*, illustrated by Michael Wolgemut (1434–1519) and his stepson Wilhelm Pleydenwurff (*c.* 1420–1472), and possibly others of the workshop. Book Six, f. CCXXXr (of the first English edition, online at http://digicoll.library.wisc.edu/cgi/t/text/text-idx?c=nur;cc=nur;view=text;idno=nur.001.0004;rgn=div2;node=nur.001.0004%3A8.285) includes reference to the plagues of 1348, stating:

> At this time locusts and vermin passed through the sky from east to west like a thick cloud, devastating all vegetation and fruits; and after they were dispersed the stench caused a horrible pestilence …
>
> A pitiful and lamentable pestilence began in the year 1348 and endured for three years throughout the world. It resulted from the aforesaid locusts or vermin. It started in India and spread as far as England, ravaging Italy and France, and finally Germany and Hungary. The mortality was so rapid and great that barely ten persons out of every thousand survived. In some regions only about one third of the population escaped. Many cities, towns, marts and villages died out entirely and remained void.

This pestilence, which formed the framework of Boccaccio's *Decameron*, is here blamed, like other plagues and famines, on the Jews (as were certain types of discord in Thomas of Cantimpré's treatise on the beehive: Cantimpré 1997: 150, 157–65), and illustrated by a small, crude, nearly square woodcut depicting locusts catapulting downward toward its lower right corner.

Toward the end of the fifteenth century in northern Europe and Italy, botanists, stimulated by the practical needs of medicine, were beginning to study and classify plants in a universalizing way, so that their principal characteristics could be described and related to other plants. Some of these botanists' early printed books, even incunabula (printed before 1501), depict what were then categorized as "Vermes," or what we have suggested may be termed "lowly creatures," including arthropods like the scorpion. In the first printed herbal, Pseudo-Apuleius's *Herbarium* (Rome: De Lignamine, 1481–1484, f. [10?]v: https://images.metmuseum.org/CRDImages/dp/original/DP357300.jpg), there appears a woodcut illustration of a scorpion (and a serpent below), with the plantain. Agnes Arber, in her book on herbals, reproduces another with Giambattista della Porta's Paracelsan remedies for a scorpion's sting (*Phytognomonica*, 1591); flying insects are included in D'Aléchamps' illustration of "*Ornithogalum magnum*" (star of Bethlehem), in his *Historia generalis plantarum* of 1586 (Arber 1938/1986: 15, 253, 118).

Leonardo da Vinci studied the flight of birds in relation to the dynamics of a bird's expert use of air flow, as we see him so carefully observing and analyzing it over periods of many hours spent in nature—in the field, as we would now say—and communicated his findings in words and drawings in his *Treatise on Flight* (*Trattato del volo degli ucelli*, 1505, Biblioteca Reale, Turin: https://airandspace.si.edu/exhibitions/codex/codex.cfm#page-2-3). Just so, he studied the dynamics of water through his drawing tools of silverpoint, charcoal, or pen and ink, capturing and arresting its swirls and waves as they occurred, analyzing them through the actions of seeing, drawing and writing.

Leonardo's interest in the dynamics of flight extended also to insects, though we have fewer examples of them among his extant works. His acute observations of dragonfly flight are recorded in his notebooks, at fol. 1051v of the Codex Atlanticus, for example: "The net-wing flies with four wings, and when the front ones are raised, the rear ones are lowered. But each pair must in itself be sufficient to sustain the full weight." "When one goes up, the other goes down." While Leonardo made quick sketches of these findings with their dynamic mechanics geometrically analyzed in the Codex Atlanticus, his studies of the anatomy of a beetle and a dragonfly are finely detailed (Figure 8.3) (Kritsky and Mader 2010: 182).

Albrecht Dürer lived and worked primarily in Nuremberg, though he traveled twice to Italy, and also north as far as Holland. Dürer's *Madonna and Child with a Multitude of Animals* (1503; Albertina, Vienna) exhibits a nearly Orphic charm in the Madonna and Child (not inappropriately, if one recalls that Orpheus could symbolize Christ throughout the Middle Ages). Certainly, the animals have been drawn to the scene, and they include at least two large beetles, one toward each lower corner (Figure 8.4).

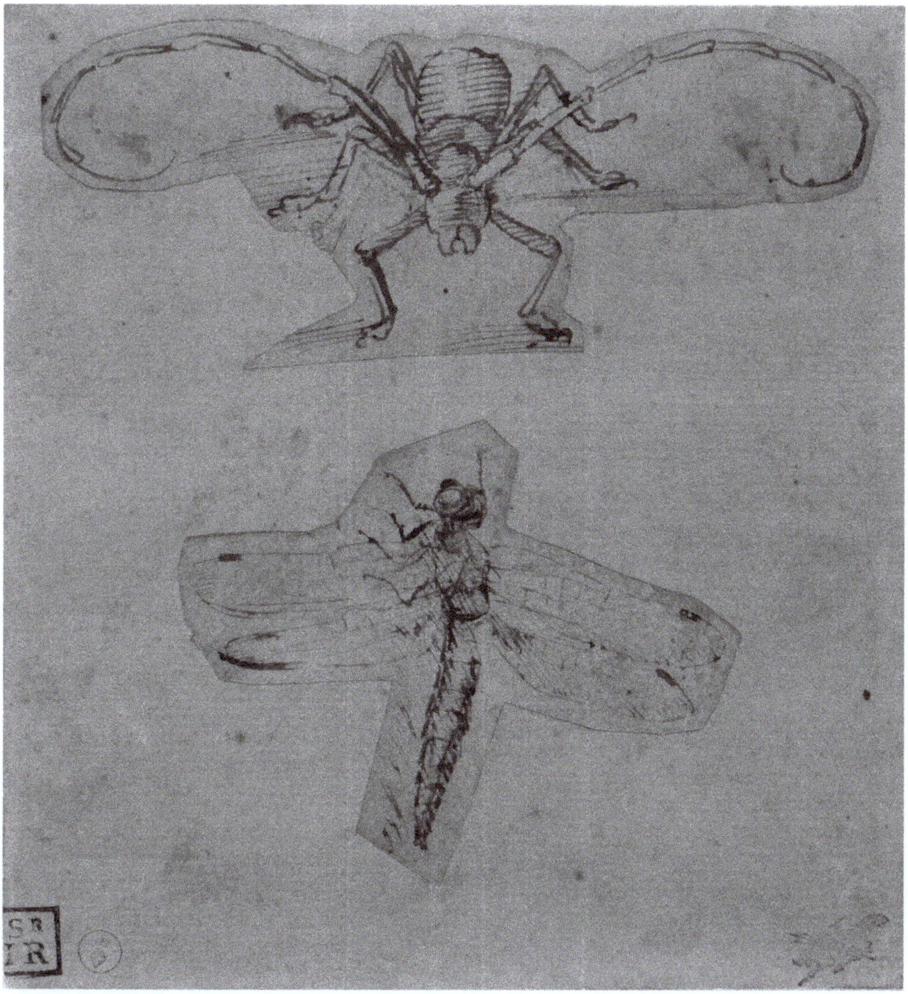

FIGURE 8.3 Leonardo da Vinci, *Longhorn Beetle and Dragonfly* (*c.* 1480 and 1503/04: Biblioteca Reale, Turin).

The attribution of the famous drawing of a stag beetle (https://books.google.com/books?id=pDMuAgAAQBAJ&printsec=frontcover#v=onepage&q&f=true) (Plate 14; pp. 287–90) to Dürer is contested by some, though it (or another version) was imitated by the Hoefnagels and Hans Hoffmann, discussed below; but the insects in this wash drawing are unmistakably in Dürer's style. Dürer's engraving of *The Holy Family and the Mayfly* (*c.* 1495), in numerous collections, clearly exhibits *Ephemera vulgata*, symbolic of the fleeting quality of earthly life,

FIGURE 8.4 Dürer, *Madonna with a Multitude of Animals* (1503; Albertina, Vienna).

in the lower right corner. Later in the century, Hoffmann's *Flowers and Beetles* (actually Hemiptera) of 1582 (Getty Museum) and *A Hare in the Forest* (1585: https://upload.wikimedia.org/wikipedia/commons/1/1d/Hans_Hoffmann_-_A_Hare_in_the_Forest_-_2001.12_-_J._Paul_Getty_Museum.jpg) including butterflies, a fly, a beetle, and other "Vermes" show his dependence on Dürer and similarity to the Hoefnagels, who will be discussed below.

Further north in 's-Hertogenbosch, Brabant, Flanders, Hieronymous Bosch (or van Aken, *c.* 1450–1516) was neither decorating prayer books nor illustrating natural history works. Instead, Bosch was allowing his fantasy full rein, to incorporate animal motifs, including those taken from insect study, into his satirical and allegorical monstrous figures. In his *Garden of Earthly Delights* (*c.* 1503), we find such metaphorical figures as the butterfly (Psyche, perhaps alluding to the souls of the departed). Insect-like creatures tumble from the heavens in both *The Last Judgment* (*c.* 1506: Akademie der bildenden Künste, Vienna) and his Workshop's *The Origins of Sin*, the leftmost panel of the *Haywain Triptych* (Figure 8.5) (1510–1516: Prado, Madrid; Fischer 2016: 122ff., 172ff., 218ff., 247ff., 250f., 258ff.).

Pieter Breugel the Elder (1525/30–1569), another Brabantian, from Breda, is considered a follower of Bosch, as we can see from the grotesque figures especially in his *Fall of the Rebel Angels* of 1562 (Royal Museums of Fine Arts, Brussels). In the center of this painting appears a pair of black and white butterfly wings flanking a head of brown hair; in the upper right corner, bees are swarming from a beehive, and the light-flooded upper left of the panel is reminiscent of Bosch's tumbling creatures in his *Last Judgment* and *The Origins of Sin*, mentioned above (Figure 8.6).

FIGURE 8.5 Jheronimus Bosch, *The Origins of Sin*, the leftmost panel of the *Haywain Triptych* (1510–1516: Prado, Madrid; detail; cf. Fischer 2016: 122ff., 172ff., 218ff., 247ff., 250f., 258ff.).

FIGURE 8.6 Pieter Bruegel the Elder, *Fall of the Rebel Angels* (1562, Royal Museums of Fine Arts, Brussels). Note bees fleeing hive in upper right corner.

SIXTEENTH-CENTURY SCIENTIFIC ILLUSTRATION

In the sixteenth century serious naturalists in England and on the Continent began to investigate the world around them and to exchange information and specimens of various plants and animals, including insects. One polymath of enormous mental capacity, Conrad Gessner of Zürich (1516–1565: philologist, bibliographer, theologian, naturalist, physician), after traveling and studying in various parts of Europe, acted as a major clearing-house of specimens of and information about animals and plants. Once re-established in his home as the town physician, he corresponded copiously with the naturalists he knew, and some he did not, in England and Europe, requesting and sharing specimens and descriptions, both verbal and drawn (see Figure 8.7).

We have noted the Flemish painters Bosch and Breugel, who incorporated insect motifs into their fantasies. The outstanding Hoefnagel family were also Flemish and endowed with fertile imagination. Joris or Georg Hoefnagel (1542–1601) was active in the Hapsburg court of Rudolf II in Vienna and Prague. This folio entitled ATER (Metropolitan Museum of Art, New York), similar to folios of the *Four Elements* manuscript (dating from about

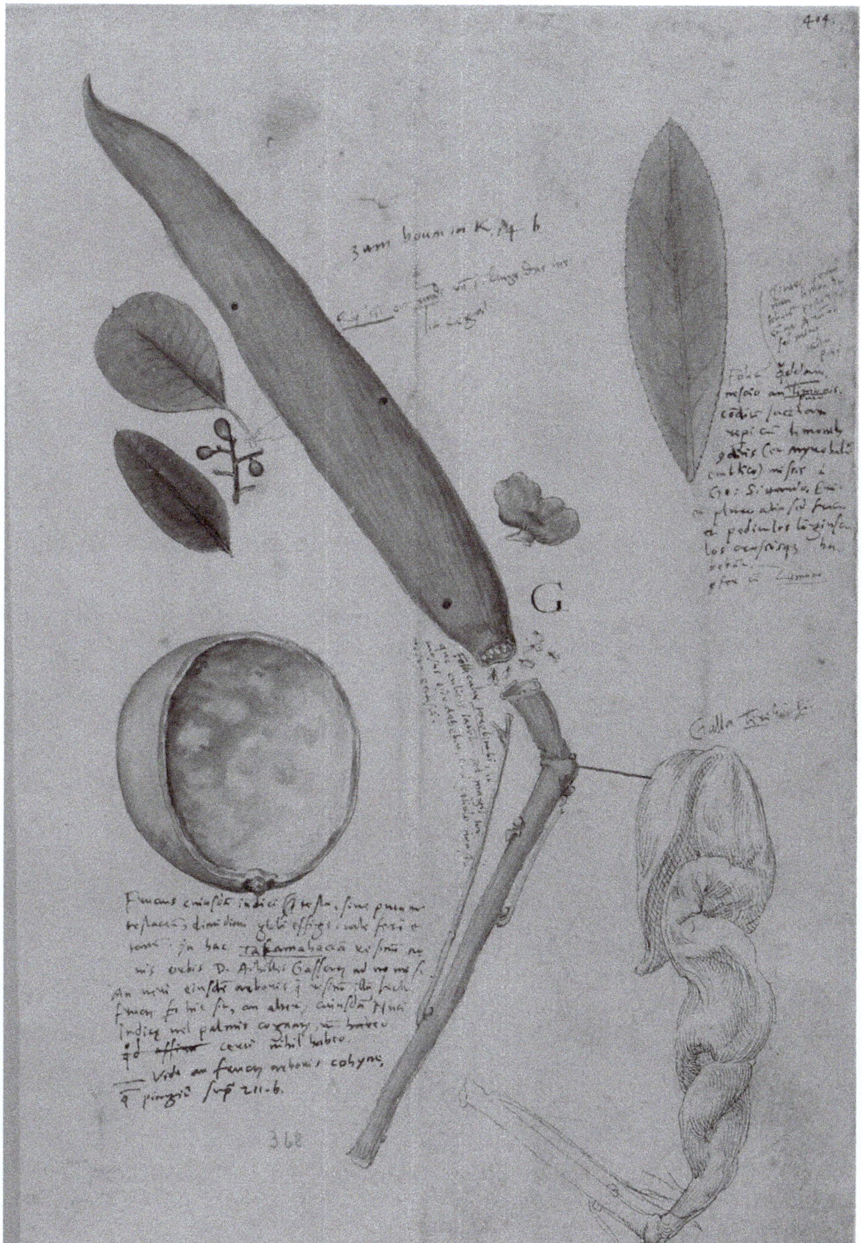

FIGURE 8.7 Conrad Gessner, *Historia Plantarum*, 1560–1565. Bib. Univ. Erlangen MS 2386.2, f. 404 (Detail). Gessner's "culices" (*Aphis pistaciae*) emerging from follicle of his "Terebinthus" (*Pistacia terebinthus* L.). Drawing made or supervised and annotated by Gessner. Photo: Holger Funk, whose transcription of the central text reads: "Folliculus terebinthi, [in] quo culices latent. Sed magis formosus esse debebat, et a petiolo non sic recte enasci."

1575–1580 and now partly dispersed, but housed mainly in the National Gallery of Art, in Washington, DC; Hendrix 1984) shows the inspired combining of mythical fantasy (the wind god) and realism (the stag beetle) in an allegorical ensemble of great aesthetic quality. This parchment folio, entitled *Ater* (Metropolitan Museum of Art, New York, discussed in a publication of the Museum (Shelley 2020) after this chapter went to press, and paired in 1989 and 2017 by Thea Vignau-Wilberg with a *Terra* in the Collection of the Princes of Liechtenstein), shows the inspired combining of mythical fantasy (the wind god) and realism (the stag beetle and other insects, some with wings appliquéd) in an allegorical ensemble of great complexity and aesthetic merit (Figure 8.8).

Earlier, we mentioned Pier Candido Decembrio's fifteenth-century treatise *On the Nature of Animals*, based on Thomas de Cantimpré's thirteenth-century *De natura rerum*. The principal manuscript, Urbinas latinus 276, of Decembrio's text is dated 1460 by Decembrio himself (Pyle 1984). Corrections of the text in his hand are visible in the margins throughout, rendering the text virtually critical, or established. The work was presented to Ludovico Gonzaga, Marquis of Mantua, who requested that Decembrio

FIGURE 8.8 Joris (Georg) Hoefnagel, *Ater* (Metropolitan Museum of Art, New York: (https://www.metmuseum.org/art/collection/search/335631). There is some question as to whether "ATER" should be read as "AÏER," though microscopic examination shows a clearly formed "T."

have a fair copy made, with space left at the base of the pages for images of the over five hundred animals, "so that I may better understand the text" (Pyle 1984, 1996a). This request—which reminds us of Frederick II's empiricism and curiosity about reality—itself defines the animal images of this manuscript (in vivid tempera colors) as illustrations, rather than decorations. Ludovico Gonzaga's request, in fact, marks the difference in priorities already alluded to between decoration and description. The two functions coexisted in the Renaissance, decorative marginalia becoming more and more accurately descriptive, though remaining decorations, while written and visual descriptions moved toward becoming scientifically more accurate, especially in the work of the sixteenth-century naturalists like the botanists Otto Brunfels (*c.* 1488–1534), Pier Andrea Mattioli (1501–1578), Leonhart Fuchs (1501–1566), and the naturalists Edward Wotton (1492–1555), Pierre Belon (1517–1564), Guillaume Rondelet (1507–1566), William Turner (1509/10–1568), Conrad Gessner and Thomas Penny (1532–1589). The final Renaissance writers on insects, Ulisse Aldrovandi (1522–1605) and Thomas Moffett (Mouffet, Muffet, 1553–1604) were sometimes more prolix than scientific (but terming them "humanists" on that account misses the point that humanist scholarship was also scientific: (Pyle 2006) but see Raven 1947: 191).

The illustrations of MS Urbinas latinus 276 have been attributed to Teodoro Ghisi (1536–1601), the Mantuan artist and naturalist who also worked with Aldrovandi (Pyle 1984, 1996a, 1996b, 2001). They illustrate this new interest in a somewhat more scientific approach to illustration, owing to the artist's use of the woodcuts from the vast tomes produced by these scientific naturalists as models, rather than woodcuts from the bestiaries or the model books which preceded them. Until he comes to describe or illustrate our "Vermes," the artist in fact relies extensively on Gessner's artists' woodcuts (some of which Gessner credits to the other naturalists mentioned, with all of whom he was in correspondence). Then, despite Ghisi's contact with Aldrovandi (whose work depended deeply on Gessner's), he clearly relies instead on his own (and possibly the Hoefnagels'; Pyle 2001) powers of observation and description in a number of cases among the "Vermes." This is famously true of his description of the economically important silkworm (featured in one edition of the *Enciclopedia Italiana*) (Figure 8.9).

As with his observation of centipedes, ants, and flies (of which there are three different species in the manuscript), Ghisi here gives full vent to his naturalist proclivities. This illustration is far more sophisticated than the woodcut in Aldrovandi's *De animalibus insectis* (1602: 282) drawn by his principal insect artist, Cornelius Schwindt, whose painted illustrations fill pages of the Bologna collection. The manuscript artist's other caterpillars too, on ff. 192v, 194v,

FIGURE 8.9 Teodoro Ghisi, *Silkworm Feeding on a Mulberry Leaf* (*Bombyx mori* on *Morus alba*, *c.* 1590. Vatican MS Urb. lat. 276, f. 197: Cf. Pyle 1984, 1996a).

and 195, are identifiable as to species, like the flies mentioned above, and other arthropods like the centipede and millipede on fol. 198.

Ghisi, though born and active in Mantua and Bologna, spent three full years, from 1587 to 1590, in Seckau and Graz at the court of the Hapsburg Archduke Charles (Karl) II (whose family had strong ties to the Gonzaga family, his sister Eleanor being Duke Guglielmo I's wife and thus Duchess of Mantua). Ghisi had worked extensively for the Gonzaga, acting as custodian of and possibly participating in the decoration of the Palazzo Te (Tellini Perina 1979: 239–68; Pyle 1996b). It is certain that he knew the ceiling paintings and the frescoes, some by Lorenzo Costa il Giovane, intimately, for their figures appear faithfully in the manuscript. So too do the woodcut outlines of Conrad Gessner's works, whose compositions are often closely imitated.

It has recently become clear (Pyle 2001) that, possibly while in Austria, Teodoro Ghisi had contact with Joris Hoefnagel, and probably also with his son Jacob (*c.* 1573–*c.* 1632; court painter to Rudolf II in Prague from 1602 to 1613). Proof of the closeness of this contact is to be found in the illustrations of a maw-worm and a leech at *Aqua*, fol. LVI of the *Four Elements* (Figure 8.10) when compared with fol. 201v of MS Urb. lat. 276. Furthermore, there appears in Aldrovandi's *De animalibus insectis* of 1602, in the woodcut illustrations at p. 765, the very same configuration, undoubtedly

FIGURE 8.10 Joris Hoefnagel, *Maw-worm and Leech* in *The Four Elements, Animalia Aqvatilia et Cochiliata (Aqva)*, f. LVI, MS (*c.* 1575/80; The National Gallery of Art, Washington, DC). Watercolor and gouache on vellum. Digitization: Rapid Capture. Image: Open Access. D-20160810-0123.jpg Gift of Mrs. Lessing J. Rosenwald. Accession No. 1987.20.7.57.

communicated to him by Ghisi, of the maw-worm, juxtaposed immediately to the left of the leech. In all three the maw-worm is draped in a similar and particularly graceful knot formation, and appears to the left of the leech. Other naturalist-artists portray segmented, rather than smooth, worms, in a different configuration, for the same description of this worm. The existence of shadows for even the most minute insect antennae is common to all three artists and will be carried over to insect drawing in seventeenth-century still-lives (Kaufmann 1993; Gombrich 2014).

While some of the insects in the manuscript are clearly drawn from life, Ghisi's principal source for his more than five hundred illustrations of MS Urb. lat. 276 was the work of the artists who filled Conrad Gessner of Zürich's many tomes with woodcut illustrations of his texts. These artists worked in close conjunction with Gessner, who himself made wash drawings of plants and insects.

Gessner's volumes on plants have been recently reproduced in a facsimile edition, which greatly extends our understanding of his modus operandi and keen eye for observation (Gessner 1987–1991). He had these characteristics in common with his English visitor in 1565, Thomas Penny, the nonconformist clergyman, physician, botanist, and entomologist who was possibly present at Gessner's death from the plague; Penny certainly collaborated with him before then and later with his heirs and executors, especially Caspar Wolf, on the preservation and dissemination of Gessner's *Nachlass*. Penny must have established a deep mutual bond with the Protestant theologian and town physician Gessner, for he was entrusted with the insect drawings which survived his lengthy return trip to England, via Montpellier, Lyon, Paris, Heidelberg, and probably Prussia. To the organization and explication of these drawings he applied his high intelligence, in view of preparing them for publication, until his own death in 1589 (Raven 1947: 153–91; cf. Funk 2018).

At this point, the drawings passed to Moffett (Raven 1947: 172–91), who assembled the drawings in a sort of scrapbook (British Library MS Sloane 4014, which offers a highly complex puzzle, under study). Moffett worked on the manuscript with Wotton, again in view of publication, adding his own observations and drawings to those of Gessner and Penny, whose classification systems (which had been increasingly based on direct observation, as opposed to tradition) he unfortunately sometimes overrode (Raven 1947: 189–91). Moffett may have succeeded in publishing a version of the text and illustrations on the Continent in 1598. After Moffett's death, Theodore Mayherne sponsored the publication, and finally succeeded in putting forth the printed *Insectorum ... Theatrum* in 1634, portions of which were published in an English version by Edward Topsell (1572–1625) (Raven 1947: 217–26; Topsell 1967).

As we have seen, the prolific Bolognese naturalist Ulisse Aldrovandi, for whom Teodoro Ghisi sometimes worked to produce animal paintings, took up the study of insects in the later sixteenth century. Much of the material in his texts derives from the Protestant Conrad Gessner's *Historiae Animalium*, which he was allowed to own by special dispensation from the archbishop of Bologna, but with the author's name censored in ink in his copies. Looking at Aldrovandi's 177 folio pages on bees (pp. 20–197, and further passages on what we now class as Hymenoptera, including wasps), we can marvel at the copiousness of the information provided. Several illustrations are noteworthy as well, (see Figure 8.11) and are claimed by Aldrovandi to be from personal observation, though that is not always certain.[1]

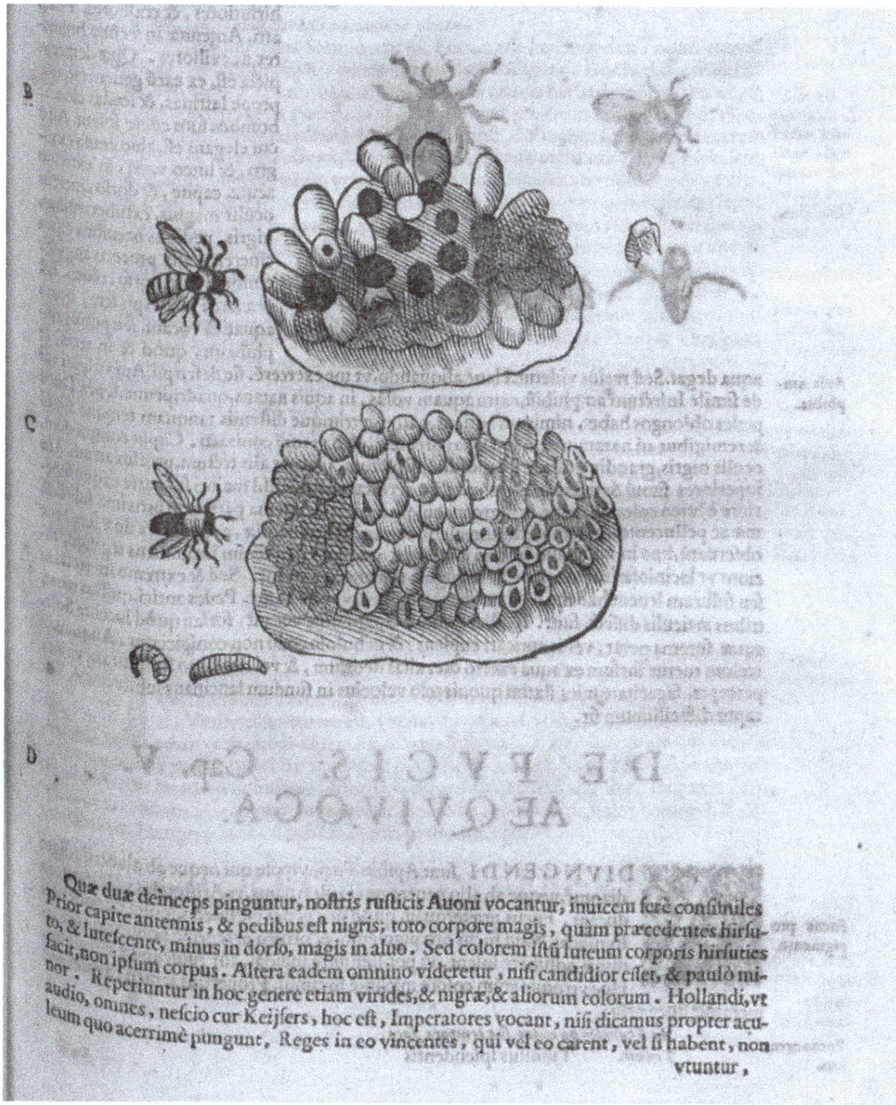

FIGURE 8.11 Aldrovandi, *De animalibus insectis*, 1602, p. 191. (Wellcome Library: https://wellcomelibrary.org/item/b17523692#?c=0&m=0&s=0&cv=206&z=-0.7182%2C-0.0766%2C2.4364%2C1.5325).

CONCURRENT METAPHORIC REPRESENTATIONS

Aldrovandi's discussion of bees, like his of all animals, includes their metaphoric and legendary roles, such as what we saw in Thomas of Cantimpré's use of the beehive as a social model for the monastery, and such as were found, in the first half of the sixteenth century, in the Emblem books of Andrea Alciati (or Alciato). Alciati depicts insects in three of his emblems: bees at number 45 (XLV. *Ex bello pax* [From war, peace]), (Figure 8.12) a fictive beetle at number 54 (LIV, *A minimis quoque timendum* [Beware of even the weakest foe]), and a scorpion at number 74 (LXXIV: *Iusta ultio* [Just revenge], showing in some editions an eagle flying off with a scorpion, which is stinging its neck with its tail).

The first is, once again, related to the positive attributes of the bee, based on the *Greek Anthology* (6, 236), by way of various Roman sources (Alciato 2009: 641–2):

> XLV
> *En galea intrepidus quam miles gesserat, & quae*
> *Saepius hostili sparsa cruore fuit.*
> *Parta pace apibus tenuis concessit in usum*
> *Alveoli, atque favos grataque mella gerit.*
> *Arma procul iaceant, fas sit tunc sumere bellum,*
> *Quando aliter pacis non potes arte frui.*
>
> (See here a helmet which a fearless soldier previously wore and which was often spattered with enemy blood. After peace was won, it retired to be used as a narrow hive for bees; it holds honey-combs and nice honey.—Let weapons lie far off; let it be right to embark on war only when you cannot in any other way enjoy the art of peace.)

Further emblematic uses of insects, bees in particular, included coats of arms, such as the family insignia of the Barberini. The Tuscan family name was originally Tafani di Barberino. "Tafano" means "horsefly," and there were three such flies on their crest, over a pair of shears referring to their wool trade. But as the family ascended socially and moved to Rome, the aggressive flies became industrious bees, later more appropriate to the dignity of a papal family; the crest was propagated throughout the city on buildings and fountains visible today.

Emblematic images certainly affected the naturalists' artists' depictions of animals. Striking examples include the smiling, humanoid-faced lions originating perhaps in early Christian and Byzantine portrayals in sculpture as well as painting, and which infiltrate the work of painters such as Mariotto Albertinelli's (1474–1515) and Alciati's lions (e.g., drawing Cupid's chariot in

EX BELLO PAX.

En galea intrepidus quam miles gesserat, & quæ
Sæpius hostili sparsa cruore fuit.
Parta pace apibus tenuis concessit in usum,
Alueoli atq; fauos grataq; melle gerit.
Arma procul iaceant, fas sit tamen sumere bellũ.
Quin aliter pacis non potes arte frui.

SVBMOVENDAM IG-
norantiam.

FIGURE 8.12 Andrea Alciato, LXV *Ex bello pax*. Augsburg, Heinrich Steyner, 1531. (From *Il libro degli Emblemi*, ed. Mino Gabriele. Milano: Adelphi, 2009, pp. 259–62.)

Emblem VII: *Potentissimus affectus Amor* [Love, the most powerful emotion]), then permeate both the painterly and the natural history traditions, occurring in Gessner's woodcuts, as well as in Teodoro Ghisi's work in MS Urb. lat. 276 at the end of the sixteenth century.

Closely entwined with the transalpine iconographical communication and interest in incorporating natural historical observations already alluded to was the remarkable work, rediscovered in the past century, of Giuseppe Arcimboldo (1526–1593). A Milanese painter, Arcimboldo was contemporary with Joris Hoefnagel and is now famous for his composite heads created from motifs of many different plants and animals, perhaps including insects (as in the work of a follower in oil on copper now in a private collection). These influential paintings were created during Arcimboldo's years in the Hapsburg imperial court of Rudolf II, where he was called in 1562 or 1563 to become the court portraitist. The reigning family had enough of a sense of humor to appreciate his exotic creations (with Rudolf II as model), and his influence grew, and has been related to the growth in the 1590s of the genre of still-life painting (Kaufmann 2009).

Another in the Flemish line was Jacob Savery (*c.* 1566–1603; also called Jacob Maertensz Saverij and Jacques Savéry). He specialized in still lifes, animals, landscapes, and genre paintings which led directly into the elaborate developments of these genres in the seventeenth century by his younger brother, Roelant Savery (or Roeland(t) Maertensz Saverij, or de Savéry, or other variants, 1576–1639). He and Ambroise Bosschaert (1573–1621) and Balthasar van der Ast (1593/94–1657) worked closely together, for a time under Abraham Bloemaert (1566–1661) in Utrecht, producing elaborate still-life paintings, which, while beginning in the 1590s, burgeoned in the seventeenth century, and will be properly dealt with in a later volume of this series.

NOTES

Introduction

1. "Renaissance" comes from the French *re* (again) and *naissance* (birth), the latter from the Latin *nascentia* (birth). It was first used by the French historian Jules Michelet in his *History of France*, Vol. II (1882).
2. See Cohen (2014) for an analysis of the changing concepts of time during the Renaissance.
3. Mann (2002) and Fagan (2019) discuss evidence for the Medieval Warm Period of the twelfth and thirteenth centuries, followed by significant cooling. Cold wet conditions in Europe, particularly in England and northern Europe, persisted to varying degrees into the fourteenth century. The second phase of the Little Ice Age saw much colder temperatures, beginning in the mid- to late 1500s, and spanned the seventeenth through nineteenth centuries in Europe (Mann 2002: 504–9).
4. The connections between *Y. pestis*, fleas (especially the rat flea, *Xenopsylla cheopis*) and rats (especially the black rat, *Rattus rattus*) were established in the late nineteenth century by Alexandre Yersin and Paul-Louis Simond. See Frith (2012) for a discussion of the history of the three great plague pandemics.
5. Gasquet (1908) gives detailed accounts of the progress of the plague across Europe, and Benedictow (2004) provides an excellent discussion of mortality data based on records from various sources, as well as a thorough examination of plague mortality in various countries.
6. It should be noted that Zeigler (1969) regarded William of Dene's estimate of "a third of the land lying idle" as an exaggeration, although it does convey the essence of the problem.
7. Farmer (2014) provides a good description of the processes involved in converting raw silk to finished threads for weaving. She also describes the roles of women in the Parisian silk industry.

8. The first known patents on inventions and processes related to silk production were granted in Venice in the 1420s; Molà (2003) provides an exhaustive list of patent applications ("charters and privileges") in Venice during the period 1437–1615.
9. The extensive use and the importance of cochineal scale pigments to the arts and to the economy of European powers, as well as in New World indigenous cultures, are discussed in Phipps (2010) and Pedilla and Anderson (2015).
10. Ogilvie (2008b) offers an excellent and comprehensive analysis of the development of the discipline of natural history during the Renaissance.
11. Bass (2019) connects Hoefnagel (and other Netherlandish artists) to the events of their time, including the Dutch Revolt of the 1560s and its impact on artistic and intellectual communities, in her thorough and detailed discussion of Hoefnagel's *Four Elements* and other works.
12. Other collectors from all over Europe contributed specimens as well; *Theatrum insectorum* was the product of a large group of natural history enthusiasts' fascination with insects and willingness to share their specimens and images of specimens. Moffet died before *Theatrum insectorum* could be published, and the manuscript eventually came into the possession of a French physician, Theodore Mayerne, who arranged for its publication in 1634.
13. See Neri (2011) and Bass (2019) for illuminating and comprehensive analyses of the ways in which art and artist-natural historians helped to shape the nature of scientific inquiry in the sixteenth century, using insects as one focus of that inquiry.

Chapter 1

1. Other translations use "gnat" in place of "flea" here. In either case, Pliny effectively poses the questions of minuteness and functional structures in insects.

Chapter 2

1. John Donne, *Poems* (1633). It is thought that he wrote this poem as a young man in the 1590s.
2. Nutton (2008) gives an insightful review of the historical controversies surrounding causes and spread of plague, and Benedictow (2004, 2010, 2016) and Little (2011) provide more extensive and comprehensive discussion of these controversies.
3. Sussman (2011) argues that it did not, having found no evidence of plague in the Yuan Dynasty or the Delhi Sultanate. Similarly, Slavin (2019) found no records of major outbreaks in Asia with the exception of an outbreak in the Chu Valley in the decade prior to the Black Death in Europe.
4. See Simond et al. (1998) for the fascinating account of Paul-Louis Simond's work on the one-hundredth anniversary of its publication.
5. See Zeigler (2016b) for an interesting and comprehensive discussion of the human flea and its potential as a vector.

6. There are numerous species of *Plasmodium* that are found in birds (avian malarias) as well as in various mammals, mostly primates.
7. Columbus's first landing site has been debated, based on interpretation of his journal and also on direction of prevailing winds, location of harbors and reefs, and other factors. Sealey (1991) provides a useful alternate interpretation.
8. There are over twenty species of *Leishmania* in South America, with possibly more that have not been described.
9. The parasites may also be acquired by eating contaminated meat, and today, by contaminated blood transfusion or organ transplant. It can also be transmitted transplacentally.

Chapter 3

1. Note that since Nahuatl was not a written language prior to the Spaniards, spellings of Nahuatl terms are not always consistent.

Chapter 4

1. Antheraea and Philosamia are genus names. Philosamia refers to *Philosamia ricini,* now *Samia cynthia ricini,* the eri silk moth.
2. The Mughal Empire, established in 1526 by the Timurid Dynasty, included the Indian subcontinent and part of central Asia. Its rulers claimed descent from Genghis Khan.
3. Warp is the lengthwise run of threads and weft refers to the crosswise threads.
4. Campbell (2002) gives a comprehensive and detailed account of the tapestries produced in Renaissance Europe, along with how they were commissioned, designed, and used from artistic, social, political, and religious perspectives.
5. The distribution of the American cochineal extended from Mexico along the western edge of South America to Cordoba, Brazil (Phipps 2012: 13).
6. Kerpel (2015: 42–3) notes that the leaders of the Indigenous Nahua people prohibited cultivation of cochineal by the 1550s since farmers had stopped growing food crops in favor of cochineal for trade to the Spanish.
7. Phipps (2015) notes that Italian guild regulations in the early sixteenth century prohibited the use of the American cochineal, but later its use was permitted, although strictly regulated, for use on velvet, damask, or satin. Cochineal dyes could not, however, be mixed; dye had to come from a single species of scale insect.
8. American cochineal could produce pinks and rose hues as well as red pigments. It became so popular that cartels grew up, controlling the supply in Europe (Greenfield 2005: 77–8).
9. Sethi (2015) notes that Polish, Armenian, kermes, and lac scales had been used in southeast Asia, China, Japan, and India for centuries prior to the introduction of American cochineal in the sixteenth century.

Chapter 5

1. The Renaissance can be understood as a transitional period in western history between a Christian society based on the authority of scripture, and a modern one based on the authority of human reason. As products of a liminal or "in-between" epoch, Renaissance inquiries into the nature of things combined older notions derived from both the Bible and the philosophers of Greco-Roman antiquity, with new insights gleaned from close observations of the natural world that anticipate the modern turn to scientific inquiry.
2. The relatively late appearance of scientific inquiry in the Renaissance period (exemplified most famously by the publication of Copernicus's heliocentric theory in the mid-1500s) can be explained in large part by the fact that the main intellectual currents of the era addressed the nature of the human being, rather than the natural world per se. The pioneering thinkers of the age are accordingly known as humanists, who found in the pagan texts of late antiquity a fresh approach to analyzing contemporary social problems, which ranged from class tensions to corruption in the Church. It is customary to mark the beginning of the Renaissance with the translations into Italian of Cicero's letters by the Tuscan lawyer Francisco Petrarca, or Petrarch (1304–1374), in the early 1300s, during a time when western Europe was weakened by plague and a schism within Christendom marked by the existence of two popes. Petrarch's publications catalyzed the search for further classical texts, a cultural demand that was happily supplied by the Italian merchant states of Venice and Florence. Italy continued to be the hub of Renaissance thought for the ensuing three centuries, receiving a fresh infusion of classical knowledge after 1453, as Byzantine Christians began to arrive, in flight from invading Ottoman forces.
3. It is important to note that the Renaissance approaches to the study of insect life, while signaling an important break from the Christian past, nevertheless stop short of constituting modern scientific exploration. Even Aldrovandi and Moffet, whose works come closest to this last mode of inquiry, ultimately defend their enterprise in metaphysical terms: in contemplating the wonder of insects, human beings are afforded an opportunity to consider the greater wonder of God.
4. Exodus 10:1–20.
5. It was through an anonymous ninth-century manuscript entitled the *Romulus Ordinarius* that eighty-three of Aesop's fables first entered into medieval European culture. In the twelfth century, Marie de France (1160–1215) translated forty of these tales from Latin into French for inclusion in her anthology of 102 "Ysopet" ("Little Aesop"), a collection of fables drawn from both classical and local sources composed in poetic form.
6. Exodus 21:28 declares: "If a bull gores a man or woman to death, the bull is to be stoned to death, and its meat must not be eaten. But the owner of the bull will not be held responsible."
7. As a general term, Christian humanism refers to an anthropology emphasizing the likeness of the human being to God, as articulated in the biblical claim that "God created mankind in his own image, in the image of God he created them; male and female he created them" (Genesis 1:27). Christian humanism during the Renaissance underscored the inherent value of the human being, while elevating the importance of human reason in creating a society reflective of Christian ideals.

8. A famous example of this approach to Christianizing paganism is recorded in the letter written by Pope Gregory I (540–604) to the Abbot Melitus, who was sent from Rome to evangelize the Anglo-Saxon people of England. As transcribed in Bede's *Ecclesiastical History of the English People* (*c.* 731), the pope's instructions are "that the temples of the idols in that nation ought not to be destroyed; but let the idols that are in them be destroyed; let water be consecrated and sprinkled in the said temples, let altars be erected, and relics placed there. For if those temples are well built, it is requisite that they be converted from the worship of devils to the service of the true God; that the nation, seeing that their temples are not destroyed, may remove error from their hearts, and knowing and adoring the true God, may the more freely resort to the places to which they have been accustomed."
9. This line of argumentation was famously advanced by sociologist Peter Berger in *The Sacred Canopy: Elements of a Sociology of Religion* (New York: Anchor Books/Doubleday, 1967). "A sky empty of angels," Berger wrote, "becomes open to the intervention of the astronomer and, eventually, of the astronaut. It may be maintained, then, that Protestantism served as a historically decisive prelude to secularization, whatever may have been the importance of other factors" (113).
10. The J. Paul Getty Museum, "Stag Beetle," http://www.getty.edu/art/collection/objects/25/albrecht-durer-stag-beetle-german-1505/ (accessed April 18, 2017). The quote comes from Dürer's *Four Books of Proportion*, translated thusly: "For Art standeth firmly fixed in Nature, and whoso can rend her forth thence, he only possesses her" (Conway 1958: 247).
11. Conway's translation reads, "Wherefore nevermore imagine that thou either canst or shalt make anything better than God hath given power to his creatures to do. For thy power is weakness compared to God's creating hand" (1958: 247).
12. The quote is widely and inaccurately attributed to Dürer himself.
13. Ambrose of Milan, *Treatise on St. Luke's Gospel,* ed. Dom G. Tissot (Paris: Cerf, 1976), 2: 193–4, qtd.in Cambefort (2006: 203).
14. For a record of modern recreations in laboratory settings of alchemical recipes, see Lawrence M. Principe, *The Secrets of Alchemy* (Chicago and London: The University of Chicago Press, 2012). The classic work on the psychological symbolism of alchemy remains Carl Jung, *Psychology and Alchemy* (Princeton, NJ: Princeton University Press, 1980).
15. The arts of alchemy were practiced during the Middle Ages for several centuries before the Renaissance, having been transmitted in twelfth-century Spain from the Islamic world into Christendom. But it was not until the 1400s that humanists in Italy rediscovered a number of texts from the Greco-Roman world that seemed to shed light on the relevance of alchemy for the restoration of Christianity. One scholar in particular, the Catholic priest Marsillo Ficinio, translated a third-century text entitled the *Corpus Hermeticum,* which contained ideas attributed to the ancient Egyptians concerning both alchemy and angelology. Most famously, Ficino also translated the entire corpus of the works of Plato into Latin, introducing this neglected strand of Greek philosophy, in its entirety, into Christendom for the first time in its history. The Florentine priest then proceeded to synthesize metaphysical ideas from both of these sources into an unprecedented Christian theology, one that he indeed hoped would enable the Church to

integrate the era's discordant speculations on the nature of God, human beings, and the natural world into a unified whole.

Ficino's synthesis, which spawned many further innovations, encompassed much more than alchemy per se, and the changes it wrought to the ancient art were profound. If ancient alchemy's goal to perfect Creation were not ambitious enough already, Renaissance or hermetic alchemy promised to confer godlike omniscience to the practitioner. In addition to making gold and achieving a state of immortality, alchemists after the era of Ficino hoped to understand the interrelationships between all fields of knowledge, gain control over the angelic powers of invisible realms, and to experience God directly. In short, the goals of this new pursuit were to know what God knows, and to know God.

16. Athanasius Kircher 1636, *Prodromus Coptus Sive Aegyptiacus* (in Cambefort 2006: 216).
17. Two titles reflective of such a historiographical approach are Joscelyn Godwin's *Athanasius Kircher: A Renaissance Man and the Quest for Lost Knowledge* (London: Thames and Hudson Ltd, 1979) and Paula Findlen's edited volume *Athanasius Kircher: The Last Man Who Knew Everything* (New York: Routledge, 2004).

Chapter 7

1. The most notable are at the Cathedral of our Lady of Strasbourg, St. Peter's Cathedral in Worms, Basel Cathedral, and at St. Sebaldus in Nuremberg.
2. In literary studies, the medieval period ranges from the sixth century through the fifteenth century. Due to that, in this chapter the term "medieval literature" will be used to refer to texts written up through the fifteenth century. Later works will be labeled "Renaissance literature."
3. *The Legend of Good Women* is a "dream vision," as are *The Parliament of Fowls* and *The House of Fame.* The dream vision was a popular literary device in the Middle Ages in which the author described a dream experienced firsthand. Often, dream visions granted the dreamer, and subsequently the reader of the poem, insight into something that might not be easily understood during waking hours.
4. When Dante sees the Uncommitted (those who in life would not pick a side), they are woefully lamenting, with blood streaming down their checks from the assault by the insects:

> Questi sciaurati, che mai non fur vivi,
> erano ignudi e stimolati molto
> da mosconi e da vespe ch'eran ivi.
>
> (*Inferno,* Canto 3, ll. 64–6)
>
> (These wretches, who never lived, were naked and bitten/stung by horseflies and wasps that were there.)

The Italian *mosconi* translates to "flies," which fits with the behavior and description of horseflies more so than hornets, as some translations read.

5. Medieval illuminated manuscripts are famous for their unique marginalia. For example, there are many instances in the illustrations of countless manuscripts that show knights battling against bunnies or gigantic snails.

6. The description provided by the British Library for this manuscript page misidentifies the insect as a "spider." Upon closer examination it can be discerned easily that the insect depicted has wings and a lot fewer legs than a spider would have.
7. During the Middle Ages, the term *fly* referred to any winged insect, including butterflies, bees, gnats, locusts, moths, etc. (Bright 1902: 110). Nevertheless, scholarly consensus on the translation has always associated the *flyyes* mentioned here with butterflies.
8. Due to the focus of this volume, only plays believed to be written before 1600 are included. For a discussion of insects in Shakespeare's plays written post-1600, see Volume 4, *A Cultural History of Insects in the Age of Enlightenment* (1600–1820).
9. Shakespeare's acting company was famously questioned after a commissioned performance of *Richard II* in 1601, a day before the rebellion against Queen Elizabeth I led by Robert Devereux, Earl of Essex. The play's focus is on the removal of a monarch from the throne. Essex was later beheaded for treason.
10. The "fly-killing scene" in *Titus Andronicus* appears in the Folio text but not in the Quarto text. Editors speculate that this likely shows that Shakespeare added the scene during a later revision of the play. Most modern editions include the scene; however, the editors of the *Norton Shakespeare* have chosen to leave the scene out of their edition. For this reason, I have referred to *The Complete Pelican Shakespeare* when quoting lines for this passage.
11. Although the poem was first published posthumously in 1633, due to the nature of the poem's topic it is likely a product of his earlier writings. Donne's work was initially either very sensual or extremely satiric; it was only much later in life that his writings took on a religious turn.

Chapter 8

1. However, it should be noted that the artist is already dissecting caterpillars: see Fig. 2.18 in Neri 2011: 72.

BIBLIOGRAPHY

Aberth, J. [2005] 2017. *The Black Death: The Great Mortality of 1348–1350, A Brief History with Documents*. Boston: Bedford/St. Martin's.

Abū al-Faẓl ibn Mubārak [1551–1602] 1873. *The Ain I Akbari*. Trans. H. Blochmann. Calcutta: G. M. Rouse.

Acosta, J. de. 2002. *Natural and Moral History of the Indies*. Trans. and ed. Jane E. Mangan. Durham, NC: Duke University Press.

Adams, M., Alther, W., Kessler, M., Kluge, M., and Hamburger, M. 2011. "Malaria in the Renaissance: Remedies From European Herbals from the 16th and 17th Century." *Journal of Ethnopharmacol* 133(2): 278–88.

Africanus, L. 1600. *A geographical historie of Africa, written in Arabicke and Italian by Iohn Leo a More, borne in Granada, and brought vp in Barbarie*. Trans. John Pory. London: George Bishop.

Africanus, L. [1600] 1896. *The History and Description of Africa: And of the Notable Things Therein Contained*. London: Bedford Press for The Hakluyt Society.

Akhoundi, M., Kuhls, K., Cannet, A., Votýpka, J., Marty, P., Delaunay, P., and Sereno, D. 2016. "A Historical Overview of the Classification, Evolution, and Dispersion of *Leishmania* Parasites and Sandflies." *PLoS Neglect Tropical Diseases* 10: e0004349.

Albee, S. and Leighton, R. 2014. *Bugged: How Insects Changed History*. New York: Bloomsbury USA Children's.

Alchon, S.A. 2003. *A Pest in the Land: New World Epidemics in a Global Perspective*. Albuquerque, NM: University of New Mexico Press.

Alciato, A. [1531] 1534. *Emblematum Liber*. In *Alciato's Book of Emblems, The Memorial Web Edition in Latin and English*. Memorial University of Newfoundland. Available online: http://www.mun.ca/alciato/index.html (accessed June 12, 2018).

Alciato, A. 2009. *Il libro degli emblemi secondo le edizioni del 1531 e del 1534*. Ed. Mino Gabriele. Milan: Adelphi. (And cf. https://www.emblems.arts.gla.ac.uk/alciato/ (accessed March 31, 2023).)

Aldrovandi, U. 1602. *De animalibus insectis libri septem, cum singulorum iconibus ad viuum expressis. Autore Vlysse Aldrouando in almo Gymnasio Bonon: ... Cum indice copiosissimo*. Bologna: apud Ioan. Bapt. Bellagambam.

Allestree, R. 1695. *The Whole Duty of Mourning and the Great Concern of Preparing Ourselves for Death*. London, printed for J. Bach.
Appleby, A. 1980. "Epidemics and Famine in the Little Ice Age." *Journal of Interdisciplinary History* 10(4): 643–63.
Appleby, J. 2014. *Shores of Knowledge: New World Discoveries and the Scientific Imagination*. New York: W.W. Norton & Company.
Arber, A. 1938/86. *Herbals. Their Origin and Evolution. A Chapter in the History of Botany 1470–1670*. Cambridge: Cambridge University Press.
Arellano, J. 2015. *Magical Realism and the History of the Emotions in Latin America*. Lanham, MD: Bucknell University Press.
Ari, T.B., Neerinckx, S., Gage, K.L., Kreppel, K., Laudisoit, A., Leirs, H., and Stenseth, N.C., 2011. "Plague and Climate: Scales Matter." *PLoS Pathogens* 7(9): p.e1002160.
Auguste, A.J., Lemey, P., Pybus, O.G., Suchard, M.A., Salas, R.A., Adesiyun, A.A., Barrett, A.D., Tesh, R.B., Weaver, S.C., and Carrington, C.V.F. 2010. "Yellow Fever Virus Maintenance in Trinidad and its Dispersal throughout the Americas." *Journal of Virology* 84(19): 9977. DOI: 10.1128/JVI.00588-10.
Badiaga, S. and Brouqui, P. 2012. "Human Louse-Transmitted Infectious Diseases." *Clinical Microbiology and Infection* 18(4): 332–7.
Bailey, M. 1988. "The Rabbit and the Medieval East Anglian Economy." *Agricultural History Review* 36(1): 1–20.
Bain, A. 2004. "Irritating Intimates: The Archaeoentomology of Lice, Fleas, and Bedbugs." *Northeast Historical Archaeology* 33(1): 81–90.
Barber, R. 1999. *Bestiary MS Bodley 764*. Woodbridge: Boydell Press.
Barragán, A.R., Dangles, O., Cardenas, R.E., and Onore, G. 2009. "The History of Entomology in Ecuador." *Annales de la Société Entomologique de France (N.S.)* 45(4): 410–23.
Bass, M. A. 2019. *Insect Artifice: Nature and Art in the Dutch Revolt*. Princeton, MA: Princeton University Press.
Battles, P., ed. 2012. *Sir Gawain and the Green Knight*. Toronto: Broadview Press.
Beagon, M. 2014. "Peter Martyr's Use of Pliny." *International Journal of the Classical Tradition* 21: 223–44.
Beavis, I.C. 1988. *Insects and Other Invertebrates in Classical Antiquity*. Liverpool: Liverpool University Press.
Bechah, Y., Capo, C., Mege, J.L. and Raoult, D. 2008. "Epidemic typhus." *The Lancet Infectious Diseases* 8(7): 417–26.
Beding, S.A., ed. 2016. *The Christopher Columbus Encyclopedia*. New York: Springer.
Benedictow, O.J. 2004. *The Black Death 1346–1353, The Complete History*. Woodbridge, Suffolk, UK: The Boydell Press.
Benedictow, O.J. 2010. *What Disease Was Plague? On the Controversy Over the Microbiological Identity of Plague Epidemics of the Past*. Leiden: Brill.
Benedictow, O.J. 2016. *The Black Death and Later Plague Epidemics in the Scandinavian Countries: Perspectives and Controversies*. Berlin: Walter de Gruyter GmbH & Co KG.
Bensi, P. 2009. "Aspects of Dyeing Techniques and Materials in Italy during the 15th and Early 16th Century." In C. Buss (ed.), *Silk Gold Crimson: Secrets and Technology at the Visconti and Sforza Courts,* 36–41. Cinisello Balsamo (Milano): Silvana.
Berdan, F. and Anawalt, P.R. 1997. *The Essential Codex Mendoza*, vol. 2. Oakland, CA: University of California Press.

Berenbaum, M. 1995. *Bugs in the System: Insects and Their Impact on Human Affairs.* New York: Helix Books.

Berenbaum, M. 2006. "On the Lives of Insects in Literature." In E.C. Brown (ed.), *Insect Poetics*, 3–12. Minneapolis/London: University of Minnesota Press.

Berman, P.S. 1994. "Rats, Pigs and Statues on Trial: The Creation of Cultural Narratives in the Prosecution of Animals and Inanimate Objects." *New York University Law Review* 69: 288–326.

Bianco, B., Alexander, R.T., and Rayson, G. 2017. "Beekeeping Practices in Modern and Ancient Yucatán. Going from the Known to the Unknown." In J.P. Mathews and T.H. Guderjan (eds.), *The Value of Things: Prehistoric to Contemporary Commodities in the Maya Region*, 87–103. Tucson, AZ: University of Arizona Press.

Bianucci, V., Giuffra, V., Bachmeier, B.E., Ball, M., Pusch, C.M., Fornaciari, G., and Nerlich, A.G. 2012. "Eleonora of Toledo (1522–1562): Evidence for Tuberculosis and Leishmaniasis Co-infection in Renaissance Italy." *Journal of Paleopathology* 2(4): 231–5.

Bitam, I., Dittmar, K., Parola, P., Whiting, M.F., and Raoult, D. 2010. "Fleas and Flea-Borne Diseases." *International Journal of Infectious Diseases* 14(8): e667–e676.

Bodenheimer, F.S. 1951. *Insects as Human Food: A Chapter of the Ecology of Man.* The Hague: Dr W. Junk Publisher.

Bos, K.I., Schuenemann, V.J., Golding, G.B., Burbano, H.A., Waglechner, N., Coombes, B.K., McPhee, J.B., DeWitte, S.N., Meyer, M., Schmedes, S., Wood, J., Earn. D.J.D., Herring, A.D., Bauer, P., Poinar, H.N., and Krause, J. 2011. "A Draft Genome of *Yersinia Pestis* From Victims of the Black Death." *Nature* 478(7370): 506–10.

Brand, D. 1988. "The Honey Bee in New Spain and Mexico." *Journal of Cultural Geography* 9(1): 71–82.

Brewer, W.H. and Farquhar, F.P. [1930] 2003. *Up and Down California in 1860–1864: The Journal of William H. Brewer*, 4th edn with maps. Oakland, CA: University of California Press. Available online: https://www.yosemite.ca.us/library/up_and_down_california/4-4.html (accessed March 31, 2023).

Bright, J.W. 1902. "Chaucer's 'bees.'" *Modern Language Notes* 37(4): 110–11.

Brittenham, C. 2015. "Three Reds: Cochineal, Hematite, and Cinnabar in the Pre-Hispanic Mesoamerican World." In C. Padilla and B. Anderson (eds.), *A Red Like No Other: How Cochineal Colored the World*, 26–35. New York: Skira Rizzoli.

Brown, E.C. 2006. "Performing Insects in Shakespeare's *Coriolanus*." In E.C. Brown (ed.), *Insect Poetics*, 29–57. Minneapolis/London: University of Minnesota Press.

Brown, J.C. 1986. "A Woman's Place Was in the Home: Women's Work in Renaissance Tuscany." In M.W. Ferguson, M. Quilligan, and N. Vickers (eds.), *Rewriting the Renaissance: The Discourses of Sexual Difference in Early Modern Europe*, 206–24. Chicago: University of Chicago Press.

Brown, M. 2015. "The Color of Power: Red in the Portraiture of the Spanish and British Empires." In C. Padilla and B. Anderson (eds.), *A Red Like No Other: How Cochineal Colored the World*, 166–73. New York: Skira Rizzoli.

Browne, T. [1672] 2009. *Pseudodoxia Epidemica: Or, Enquiries Into Commonly Presumed Truths*. Benediction Books.

Bruquetas, R. and Gómez, M. 2015. "Carmine of the Indies: Cochineal in Spanish Painting and Sculpture (1550–1670)." In C. Padilla and B. Anderson (eds.), *A Red Like No Other: How Cochineal Colored the World*, 190–201. New York: Skira Rizzoli.

Burnham, D.K. 1980. *Warp and Weft: A Textile Terminology*. Ontario: Royal Ontario Museum.
Burr, M. 1938. "Unpublished for Over Three Hundred Years. The Original Title Page for the First Book on Natural History Printed in England." *The Field*, August 27, 495.
Butler, C. 1609. *The Feminine Monarchie or a Treatise Concerning Bees*. Oxford: Joseph Barnes.
Bryant, J.E., Holmes, E.C., and Barrett, A.D.T. 2007. "Out of Africa: A Molecular Perspective on the Introduction of Yellow Fever Virus into the Americas." *PLoS Pathogens* 3(5). DOI: 10.1371/journal.ppat.0030075.
Buckland, P. 1988. "North American Faunal Connections—Introduction or Endemics?" *Entomological Scandinavia Supplement* 32: 8–29.
Burns, J., Acuna-Soto, R., and Stahle, D. 2014. "Drought and Epidemic Typhus, Central Mexico, 1655–1918." *Emerging Infectious Diseases* 20(3): 442–7.
Bussotti, M. and Qi, H. 2014. "Typography for a Modern World? The Ways of Chinese Movable Types." *East Asian Science, Technology, and Medicine* 40: 9–44.
Busvine, J. 1976. *Insects, Hygiene and History*. London: The Athlone Press.
Butazzi, G. 2009. "Some Considerations on Late-15th-Century Fashions at the Sforza Court." In C. Buss (ed.), *Silk Gold Crimson: Secrets and Technology at the Visconti and Sforza Courts*, 25–9. Cinisello Balsamo (Milano): Silvana.
Butler, C. [1609] 1623. *The Feminine Monarchie: Or the Historie of Bees*. London: John Haviland.
Buss, C. 2009. "Silk Gold Crimson." In C. Buss (ed.), *Silk Gold Crimson: Secrets and Technology at the Visconti and Sforza Courts*, 44–62. Cinisello Balsamo (Milano): Silvana.
Cambefort, Y. 1994. "Beetles as Religious Symbols." *Cultural Entomology Digest* 2: 1–14.
Cambefort, Y. 2006. "A Sacred Insect on the Margins: Emblematic Beetles in the Renaissance." In E.C. Brown (ed.), *Insect Poetics*, 200–22. Minneapolis/London: University of Minnesota Press.
Camerarius, J. 1596. *Symbolorum et Emblematum ex volatilibus et insectis*. Nuremberg: Vogelin.
Campbell, B.M.S. 2016. *The Great Transition: Climate, Disease and Society in the Late-Medieval World*. Cambridge: Cambridge University Press.
Campbell, T.P. 2002. *Tapestry in the Renaissance: Art and Magnificence*. New York: The Metropolitan Museum of Art.
Campbell, Thomas P. 2008. "How Medieval and Renaissance Tapestries Were Made." In *Heilbrunn Timeline of Art History* (February 2008). New York: The Metropolitan Museum of Art, 2000. Available online: http://www.metmuseum.org/toah/hd/tapm/hd_tapm.htm (accessed March 31, 2023).
Camuffo, D. and Enzi, S. 1991. "Locust Invasions and Climatic Factors from the Middle Ages to 1800." *Theoretical and Applied Climatology* 43(1–2): 43–73.
Cantimpré, T. de. 1973. *Liber de natura rerum*, Teil I, Text. Ed. H. Boese. Berlin and New York: W. De Gruyter.
Cantimpré, T. de. 1997. *Les exemples du Livre des Abeilles*. Ed. H. Platelle. Tournhout: Brepols.
Cantor, N.F. 2001. *In the Wake of the Plague*. New York: The Free Press.
Charles-Edwards, T. and Kelly, F., trans. 1983. *Bechbretha: [An Old Irish Law-Tract on Beekeeping]*. Dublin: Dublin Institute for Advanced Study.

Charlier, P., Bouchet, F., Weil, R., and Bonnet, B. 2015. "Schistosomiasis in the Mummified Viscera of Saint-Louis (1270 AD)." *Forensic Science, Medicine, and Pathology*, (Oct.) 1–2. Available online: http://doi.org/10.1007/s12024-015-9722-4 (accessed March 31, 2023).

Chastel, A. 1984. *A Chronicle of Italian Renaissance Painting*. New York: Cornell University Press.

Chaucer, G. [1387–1400] 1987. *The Canterbury Tales*. In L.D. Benson, *The Riverside Chaucer*. Boston: Houghton Mifflin Company.

Chaucer, G. [1379–1380] 1987. *The House of Fame*. In L.D. Benson, *The Riverside Chaucer*. Boston: Houghton Mifflin Company.

Chaucer, G. [1379–1380] 1987. *The Legend of Good Women*. In L.D. Benson, *The Riverside Chaucer*. Boston: Houghton Mifflin Company.

Chaucer, G. [1379–1380] 1987. *The Parliament of Fowles*. In L.D. Benson, *The Riverside Chaucer*. Boston: Houghton Mifflin Company.

Chidester, D. 2001. *Christianity: A Global History*. New York: HarperOne.

Cholewiński, M., Derda, M., and Hadaś, E. 2017. "Hygiene Pests as Vectors for Parasitic and Bacterial Diseases in Humans." *Annals of Parasitology* 63(2): 81–97.

Christodoulopoulos, G., Theodoropoulos, G., Kominakis, A., and Theis, J.H. 2006. "Biological, Seasonal and Environmental Factors Associated with *Pulex irritans:* Infestation of Dairy Goats in Greece." *Veterinary Parasitology* 137(1–2): 137–43.

Clarke, M., ed. 2016. *The crafte of lymmyng and the maner of steynyng*. Oxford: Oxford University Press.

Cobb, M. 2000. "Reading and Writing the Book of Nature: Jan Swammerdam (1637–1680)." *Endeavour* 24: 122–8.

Cobe, J.V. 1499. *Hortus Sanitatis*. Straßburg: Johann Prüss.

Cohen, S. 2014. *Transformations of Time and Temporality in Medieval and Renaissance Art*. Leiden: Brill.

Cohn Jr., S.K. 2008. "Epidemiology of the Black Death and Successive Waves of Plague." *Medical History Supplement* 27: 74–100.

Cohn Jr., S.K. 2018. *Epidemics: Hate and Compassion from the Plague of Athens to AIDS*. Oxford: Oxford University Press.

Cole, L. 2010. "Of Mice and Moisture: Rats, Witches, Miasma, and Early Modern Theories of Contagion." *Journal for Early Modern Cultural Studies* 10(2): 65–84.

Cole, L. 2016. *Imperfect Creatures: Vermin, Literature and the Science of Life 1600–1740*. Ann Arbor: University of Michigan Press.

Cole, L. 2019. "Swift Among the Locusts: Vermin, Infestation, and Natural Philosophy in the Eighteenth Century." In B. Boehrer, M. Hand, and B. Massumi (eds.), *Animals and Animality in the Literary Field*, 136–55. Cambridge: Cambridge University Press.

Connor, S. 2006. *Fly*. London: Reaktion.

Conway, W.M., ed. and trans. 1958. *The Writings of Albrecht Dürer*. New York: Philosophical Library.

Cook, N.D. 2002. "Sickness, Starvation, and Death in Early Hispaniola." *Journal of Interdisciplinary History* 32(3): 349–86.

Costa, M.A., Matheson, C., Iachetta, L., Llagostera, A., and Appenzeller, O. 2009. "Ancient Leishmaniasis in a Highland Desert of Northern Chile." *PLoS One* 4: e6983.

Costa-Neto, E.M. 2002. "The Use of Insects in Folk Medicine in the State of Bahia, Northeastern Brazil, with Notes on Insects Reported Elsewhere in Brazilian Folk Medicine." *Human Ecology* 30: 245–63.

Coutts, N. 2006. "Portraits of the Nonhuman: Visualizations of the Malevolent Insect." In E.C. Brown (ed.), *Insect Poetics*, 298–318. Minneapolis and London: University of Minnesota Press.

Crane, E. 1999. *The World History of Beekeeping and Honey Hunting*. New York: Routledge.

Crescentiis, P.V. 1518. *Opus Ruralium commodorum*. Straßburg.

Crosby Jr., A.W. 1986. *Ecological Imperialism: The Biological Expansion of Europe, 900–1900*. Cambridge: Cambridge University Press.

Culpeper, N. 1653. *Pharmacopoeia Londinensis, or The London dispensatory ... by Nich. Culpeper, Gent*. London: Peter Cole.

Curran, C.H. 1937. "Insect Lore of the Aztecs." *Natural History* 39: 196–203.

Dale, M.K. 1933. "The London Silkwomen of the Fifteenth Century." *The Economic History Review* 4(3): 324–35.

Daly, P. 2014. *The Emblem in Early Modern Europe: Contributions to the Theory of the Emblem*. Farnham: Ashgate.

Daston, L. and Park, K. 1998. *Wonders and the Order of Nature, 1150–1750*. New York: Zone Books.

Davenport, F.G. 1917. *European Treaties Bearing on the History of the United States and its Dependencies to 1648*, vol. 1. Washington, DC: Carnegie Institution of Washington.

Davies, S. 2016. *Renaissance Ethnography and the Invention of the Human: New Worlds, Maps and Monsters*. Cambridge: Cambridge University Press.

De Hamel, C. 1986. *A History of Illuminated Manuscripts*. Boston: David R. Godine.

de Landa, D. 1937. *Yucatan Before and After the Conquest (1937)*. Trans. William Gates. Available online: https://www.globalgreyebooks.com/yucatan-before-and-after-the-conquest-ebook.html (accessed April 10, 2023).

de Quatrefages, J.L.A. 1873. "Silk-Worms and Sericulture." Trans. Eliza A. Youmans. *Popular Science Monthly* 3 (October): 657–76.

de Vries, J. 1989. "Renaissance Cities." *Renaissance Quarterly* 42: 781–93.

Dean, K.R., Krauer, F., Walløe, L., Lingjærde, O.C., Bramanti, B., Stenseth, N.C., and Schmid, B.V. 2018. "Human Ectoparasites and the Spread of Plague in Europe During the Second Pandemic." *PNAS* 115(6): 1304–9.

Dean, K.R., Krauer, F., and Schmid, B.V. 2019. "Epidemiology of a Bubonic Plague Outbreak in Glasgow, Scotland in 1900." *Royal Society Open Science* 6(1): 181695. Available online: https://royalsocietypublishing.org/doi/full/10.1098/rsos.181695 (accessed March 31, 2023).

Decembrio, P.C. 1984. *De animantium naturis* (1460). Vatican MS Urbinas latinus 276. Ed. C.M. Pyle. Zürich: Belser Verlag (Codices e Vaticanis Selecti, LX). Available online: https://digi.vatlib.it/view/MSS_Urb.lat.276 (accessed March 31, 2023).

Dee, J. [1564] 2000. *The Hieroglyphic Monad*. Boston and York Beach, ME: Weiser Books.

DeFoliart, G. 2002. "The Human Use of Insects as a Food Resource." Available online: https://insectsasfood.russell.wisc.edu/the-human-use-of-insects-as-a-food-resource/ (accessed February 2, 2018).

Delaunay, P., Blanc, V., Del Giudice, P., Levy-Bencheton, A., Chosidow, O., Marty, P., and Brouqui, P. 2011. "Bedbugs and Infectious Diseases." *Clinical Infectious Diseases* 52(2): 200–10.

Dicke, M. 2000. "Insects in Western Art." *American Entomologist* 46(4): 228–37.

Dinzelbacher, P. 2002. "Animal Trials: A Multidisciplinary Approach." *Journal of Interdisciplinary History* 32(3): 405–21.

Dobson, M.J. 1989. "History of Malaria in England." *Journal of the Royal Society of Medicine* Supplement No. 17, 82: 3–7.

Dobson, M.J. 1997. *Contours of Death and Disease in Early Modern England*. Cambridge: Cambridge University Press.

Donne, J. [1633] 2011. "The Flea." In J. Black, L. Conolly, K. Flint, I. Grundy, D. LePan, R. Liuzza, J.J. McGann, A.L. Prescott, B.V. Qualls, and C. Waters (eds.), *The Broadview Anthology of British Literature*, Concise Edition, vol. A, 831. Toronto: Broadview Press.

Doran, M. 1940. "On Elizabethan 'Credulity': With Some Questions Concerning the Use of the Marvelous in Literature." *Journal of the History of Ideas* 1: 151–76.

Doutt, R.L. 1964. "The Historical Development of Biological Control." In P. DeBach (ed.), *Biological Control of Insect Pests and Weeds*, 21–42. New York: Reinhold Publishing Company.

The Drake Manuscript in the Pierpont Morgan Library: Histoire Naturelle des Indes. 1996. London: André Deutsch.

Drancourt, M., Houhamdi, L., and Raoult, D. 2006. "*Yersinia Pestis* as a Telluric, Human Ectoparasite-Borne Organism." *The Lancet Infectious Diseases* 6(4): 234–41.

Durand, J.D. 1977. "Historical Estimates of World Population: An Evaluation." *Population and Development Review* 3: 253–96.

Dyer, C. 2005. *An Age of Transition? Economy and Society in England in the Later Middle Ages*. Oxford: Oxford University Press.

Dyer, T.F. Thiselton. [1883] 2010. *Folk-lore of Shakespeare*. Whitefish, MT: Kessinger Publishing, LLC. Available online: https://www.sacred-texts.com/sks/flos/flos11.htm#fr_573 (accessed April 33, 2023).

Eads, D.A. and Biggins, D.E. 2015. "Plague Bacterium as a Transformer Species in Prairie Dogs and the Grasslands of Western North America." *Conservation Biology* 29(4): 1086–93.

Eamon, W. 2007. "The Canker Friar: Piety and Intrigue in an Era of New Diseases." In F. Mormando and T. Worcester (eds.), *Piety and Plague: From Byzantine to the Baroque*, 156–73. Kirksville, MO: Truman State University.

Earle, R. 2010. "If You Eat Their Food ...: Diets and Bodies in Early Colonial Spanish America." *The American Historical Review* 115(3): 688–713.

Ecobichon, D.J. 2001. "Toxic Effects of Pesticides." In C.D. Klaassen (ed.), *Casarett and Doull's Toxicology*, 6th edn, 763–810. New York: McGraw-Hill Publishers.

Eisen, R.J. and Gage, K.L. 2012. "Transmission of Flea-Borne Zoonotic Agents." *Annual Review of Entomology* 57: 61–82.

Eisenstein, E. 1980. *The Printing Press as an Agent of Change*. 2 vols. Cambridge: Cambridge University Press.

El Hamzaoui, B., Laroche, M., Bechah, Y., Berenger J.M., and Parola, P. 2019. Testing the Competence of *Cimex lectularius* Bed Bugs for the Transmission of *Borrelia recurrentis*, the Agent of Relapsing Fever. *American Journal of Tropical Medicine and Hygiene* 100(6): 1407–12.

Eliade, M. 1978. *The Forge and the Crucible: The Origins and Structures of Alchemy*. 2nd edn. Chicago and London: The University of Chicago Press.

Elliott, J.H. 1970. *The Old World and The New, 1492–1650*. Cambridge: Cambridge University Press.

Evans, E.P. [1906] 2009. *The Criminal Prosecution and Capital Punishment of Animals*. Reprint edn. New Jersey: Lawbook Exchange Ltd.

Evans, J., Alemu, M.H., Flore, R., Frost, M.B., Halloran, A., Jensen, A.B., Maciel-Vergara, G., Meyer-Rochow, V.B., Münke-Svendsen, C., Olsen, S.B., Payne, C., Roos, N., Rozin, P., Tan, H.S.G., van Huis, A., Vantomme, P., and Ellenberg, J. 2015. "'Entomophagy': An Evolving Terminology in Need of Review." *Journal of Insects as Food and Feed* 1(4): 293–305. Available online: https://doi.org/10.3920/JIFF2015.0074 (accessed April 10, 2023).

Ewen, C.L. 1970. *Witch Hunting and Witch Trials*. London: Fred Miller.

Fagan, Brian M. 2019. *The Little Ice Age: How Climate Made History 1300–1850*. New York: Basic Books.

Fan, C. 2011. *Treatises of the Supervisor and Guardian of the Cinnamon Sea*. Trans. James M. Hargett. Seattle: University of Washington Press.

Farmer, S. 2014. "Medieval Paris and the Mediterranean: The Evidence from the Silk Industry." *French Historical Studies* 37(3): 383–419.

Farmer, S. 2016. *The Silk Industries of Medieval Paris: Artisanal Migration, Technological Innovation, and Gendered Experience*. Philadelphia: University of Pennsylvania Press.

Figuier, L. 1872. *The Insect World: Being a Popular Account of the Orders of Insects*. New York: Cassell, Petter, and Galpin.

Findlen, P. 2004. *Anthanasius Kircher: The Last Man Who Knew Everything*. New York and London: Routledge.

Fischer, S. 2016. *Jheronimus Bosch*. Köln: Taschen.

Fissell, M. 1999. "Imagining Vermin in Early Modern England." *History Workshop Journal* 47(1): 1–29.

Fitzherbert, J. [1523] 1534. *Book of Husbandry*. Ed. W.W. Skeat, 1882. London: Trübner & Co., Ludgate Hill. Available online (1534 edition): https://www.gutenberg.org/files/57457/57457-h/57457-h.htm#S_122 (accessed April 33, 2023).

Fliegel, S.N. 2016. "A Wax Artrist's Model from the Renaissance." Available online: https://www.lagazzettaitaliana.com/history-culture/8056-a-wax-artist-s-model-from-the-renaissance (accessed April 3, 2023).

Forbes, R.J. 1956. *Studies in Ancient Technology*, vol. IV. Leiden: E.J. Brill.

Forbes, V., Dussault, F., and Bain, A. 2013. "Contributions of Ectoparasite Studies in Archaeology with Two Examples from the North Atlantic Region." *International Journal of Paleopathology* 3: 158–64.

Fornaciari, G., Giuffra, V., Marinozzi, S., Picchi, M.S. and Masetti, M. 2009. "'Royal' Pediculosis in Renaissance Italy: Lice in the Mummy of the King of Naples Ferdinand II of Aragon (1467–1496)." *Memórias do Instituto Oswaldo Cruz* 104(4): 671–2.

Fornaciari, G., Giuffra, V., Ferroglio, E., Gino, S., and Bianucci, R. 2010. "*Plasmodium falciparum* Immunodetection in Bone Remains of Members of the Renaissance Medici Family (Florence, Italy, Sixteenth Century)." *Transactions of the Royal Society of Tropical Medicine and Hygiene* 104(2010): 583–7.

Fornaciari, G., Marinozzi, S., Gazzaniga, V., Giuffra, V., Picchi, M.S., Giusiani, M., and Masetti, M. 2011. "The Use of Mercury Against Pediculosis in the Renaissance: The Case of Ferdinand II of Aragon, King of Naples, 1467–96." *Medical History* 55(1): 109–15.

Forster, K.A. 1975. *Die Biene*. Zürich: Küsnacht.

Franchini, D.A., Margonari, R., Olmi, G., Signorini, R., Zanca, A., and Perina, C.T. 1979. *La Scienza a Corte. Collezionismo eclettico, natura e immagine a Mantova fra Rinascimento e Manierismo*. Rome: Bulzoni.

Franklin, P. 1983. “Malaria in Medieval Gloucestershire: An Essay in Epidemiology.” *Transactions of the Bristol and Gloucester Archaeological Society* 101: 111–22.

Fraser, R.D.B. and T.P. MacRae. 1973. *Conformation in Fibrous Proteins and Related Synthetic Polypeptides*, Ch. 13: Silks. New York: Academic Press, Inc.

Freeman, R. 1948. *English Emblem Books*. London: Chatto and Windus.

Frith, J. 2012. The History of Plague – Part 1. The Three Great Pandemics. *J. Military and Veterans Health* 20(2). Available online: https://jmvh.org/article/the-history-of-plague-part-1-the-three-great-pandemics/ (accessed April 3, 2023).

Funk, H. 2018. “Thomas Penny and Conrad Gessner: A Revision of Some Long-Established Chronological Particulars.” *Archives of Natural History* 45: 166–9.

García, M. del C.L. 2002. *La distinción alimentaria de Toluca: el delicioso valle y los tiempos de escasez, 1750–1800*, CIESAS. Available online: https://books.google.com/books?id=CwORekx-WKUC&pgis=1 (accessed February 17, 2018).

Garcia-Herrera, E., Méndez-Gallegos, S., and Talavera-Magaña, D. 2010. “El género *Agave* spp. en México: Principales usos de importancia socioeconómica y agroecológica.” *Revista Salud Pública y Nutrición*, edición especial 5: 109–27.

Gardner, W. 1931–1933. “A Lancashire Entomologist [Thomas Penny] in the Time of Queen Elizabeth.” *Annual Reports & Proceedings of the Lancashire and Cheshire Entomological Society* 55, 56, 57: 31–52.

Garrelt, C. and Wiechmann, I. 2003. “Detection of *Yersinia Pestis* DNA in Early and Late Medieval Bavarian Burials.” In G. Grupe and J. Peters (eds.), *Decyphering Ancient Bones; The Research Potential of Bioarchaeological Collections*, 247–54. Documenta Archaeobiologiae; Bd. 1. Leidorf, Germany: Rahden/Westf.

Gasch-Tomás, J.L. 2018. “The Manila Galleon and the Reception of Chinese Silk in New Spain, c. 1550–1650.” In D. Schäfer, G. Riello, and L. Móla (eds.), *Threads of Global Desire: Silk in the Pre-Modern World*, 251–64. Rochester, NY: Boydell & Brewer.

Gasquet, F. 1908. *The Black Death of 1348 and 1349*, 2nd edn. London: George Bell and Sons.

Gerarde, J. [1582] 1597. *The Herball, or, Generall historie of plantes*. I. London: J. Norton. Available online: https://archive.org/details/mobot31753000817749 (accessed April 3, 2023).

Gerszten, E., Allison, M., and Maguire, B. 2012. “Paleopathology in South American Mummies: A Review and New Finding.” *Pathobiology* 79: 247–56.

Gessner, C. 1551–1587. *Conradi Gesneri … Historiae animalium: Liber I. De quadrupedibus viviparis* (Zurich: Froschauer, 1551); *Liber II. De quadrupedibus oviparis* (Zurich: Froschauer, 1554); *Liber III qui est de avium natura* (Zurich: Froschauer, 1555); *Liber IIII qui est de piscium & aquatilium animantium natura* (Zurich: Froschauer, 1558); *Liber V qui est de serpentium natura*. Ed. J. Carron (Zurich: Froschauer, 1587).

Gessner, C. 1587. *Historiae insectorum Libellus, qui est De Scorpione*. Tiguri: Froschoverus.

Gessner, C. 1987–1991. *Conradi Gesneri historia plantarum, Gesamtausgabe, enthaltend acht faksimilierte und 320 mehrfarbige, verkleinerte Aquarelle aus dem botanischen Nachlass von Conrad Gessner (1516–1565) in der Universitätsbibliothek Erlangen*. Eds. Heinrich Zoller and and Martin Steinmann. 2 vols. Dietikon-Zürich: Urs Graf-Verlag.

Glanvill, J. 1668. *Plus ultra, or, The progress and advancement since the days of Aristotle*. London: James Collins.

Glanvill, J. 1681. *Saducimus triumphatus, or Full and plain evidence concerning witches and apparitions in two parts. Part the Second*. London: Printed for J. Collins and S. Lownds.

Goldner, G.R. 1988. *European Drawings 1. Catalogue of the Collections*. Malibu: The J. Paul Getty Museum.

Gombrich, E.H. 2014. *Shadows: The Depiction of Cast Shadows in Western Art*. New Haven, CT: Yale University Press.

Goodman, G. 1616. *The fall of man, or the corruption of nature*. London: Felix Kingston.

Gottfried, R.S. 1983. *The Black Death: Natural and Human Disaster in Medieval Europe*. New York: The Free Press (Macmillan Publishing Co., Inc.).

Gotzek, D., Axen, H.J., Suarez, A.V., Cahan, S.H., and Shoemaker, D. 2015. "Global Invasion History of the Tropical Fire Ant: A Stowaway on the First Global Trade Routes." *Molecular Ecology* 24(2): 374–88.

Grafton, A. 1992. *New Worlds, Ancient Texts: The Power of Tradition and the Shock of Discovery*. Cambridge, MA: Harvard University Press.

Green, M.H. 2014. "Taking 'Pandemic' Seriously: Making the Black Death Global." *The Medieval Globe* 1(1): 27–61.

Green, M.H. 2018. "Putting Africa on the Black Death Map: Narratives from Genetics and History." *Afriques. Débats, méthodes et terrains d'histoire* (09). Available online: https://journals.openedition.org/afriques/2125 (accessed November 1, 2019).

Greenblatt, S. 1992. *Marvelous Possessions: The Wonder of the New World*. Chicago: University of Chicago Press.

Greenblatt, S. 2011. *The Swerve: How the World Became Modern*. New York: Norton.

Greenfield, A.B. 2005. *A Perfect Red: Empire, Espionage, and the Quest for the Color of Desire*. New York: HarperCollins.

Guellil, M., Kersten, O., Namouchi, A., Bauer, E.L., Derrick, M., Jensen, A.Ø., Stenseth, N.C., and Bramanti, B. 2018. "Genomic Blueprint of a Relapsing Fever Pathogen in 15th Century Scandinavia." *PNAS* 115(41): 10422–7.

Guhl, F. 2017. "Geographical Distribution of Chagas Disease." In J. Tellería and M. Tibnayrenc (eds.), *American Trypanosomiasis (Chagas Disease): One Hundred Years of Research*, 2nd edn, 89–112. London: Elsevier.

Gullan, P.J. and Cranston, P.S. 2014. *The Insects: An Outline of Entomology*, 5th edn. New York: Wiley & Sons.

Gunnell, N. 2012. "A Look at the Social Effects of the Black Death." Available online: http://www.brighthubeducation.com/history-homework-help/88775-social-effects-of-the-black-death/ (accessed February 20, 2018).

Haensch, S., Bianucci, R., Signoli, M., Rajerison, M. Schultz, M., Kacki, S., Vermunt, M., Weston, D., Hurst, D., Achtman, M., Carniel, E., and Bramanti, B. 2010. "Distinct Clones of *Yersinia Pestis* Caused by the Black Death." *PLoS Pathogens* 6(10): e1001134.

Harkness, D.E. 2009. "Elizabethan Naturalists and the Work of John White." *European Visions: American Voices* 172: 44–50. British Museum Research Publication.

Harris, M. 1986. *Good to Eat: Riddles of Food and Culture*. New York: Simon & Schuster.

Harrison, T.P. 1971. "Shakespeare's Glowworms." *Shakespeare Quarterly* 22(4): 395–6.

Harvard University. 2005. "Solving The Mystery Of Centuries-old Caribbean Ant Plagues." *ScienceDaily*, January 28. Available online: http://www.sciencedaily.com/releases/2005/01/050127234358.htm (accessed February 22, 2018).

Haskins, C.H. 1921. "The 'De Arte Venandi cum Avibus' of the Emperor Frederick II." *The English Historical Review* 36: 334–55.

Hays, J.N. 2005. *Epidemics and Pandemics: Their Impacts on Human History*. Santa Barbara, CA: ABC-CLIO.

Hecker, J.F.C. 1859. *The Epidemics of the Middle Ages*. Trans. B.G. Babington. London: Trübner & Co.

Hempelmann, E. and Krafts, K. 2013. "Bad Air, Amulets and Mosquitoes: 2,000 Years of Changing Perspectives on Malaria." *Malaria Journal* 12: 232. Available online: https://doi.org/10.1186/1475-2875-12-232 (accessed April 3, 2023).

Henderson, E.F. 1896. *Select Historical Documents of the Middle Ages*. London: George Bell and Sons. Available online: https://avalon.law.yale.edu/medieval/statlab.asp (accessed April 10, 2023).

Hendrix, L. 1995. "Of Hirsutes and Insects: Joris Hoefnagel and the Art of the Wondrous." *Word & Image* 2(4): 373–90.

Hendrix, M.L. 1984. "Joris Hoefnagel and the Four Elements: A Study of Sixteenth-Century Nature Painting." PhD diss., Princeton University, Ann Arbor: University Microfilms.

Herlihy, D. 1997. *The Black Death and the Transformation of the West*. Cambridge, MA: Harvard University Press.

Hernández, F. 1959. *Obras completas: pt.1–2. Historia natural de nueva España*. Universidad Nacional Autónoma de México. Available online: http://www.franciscohernandez.unam.mx/home.html (accessed September 14, 2018).

Hernández de Toledo, F., Barros, C., and Buenrostro, M. 2007. *La alimentación de los antiguos Mexicanos en la Historia natural de Nueva España de Francisco Hernández*. México, D.F., Universidad Nacional Autónoma de México.

Heydenreich, L.H. 1954. *Leonardo da Vinci*. Trans. D.J. Janson, M. Barnes et al., 2 vols. London: Allen and Unwin.

Heymsfeld, Ralph. n.d. "Life at Sea in the Time of Magellan." Available online: http://www.thepeacefulsea.com/life-at-sea.html (accessed February 22, 2018).

Heywood, J. 1556. *The spider and the flie*. London: Tho. Powell.

The Hieroglyphics of Horapollo Nilous. Available online: http://www.sacred-texts.com/egy/hh/index.htm (accessed April 3, 2023).

Hirst, K.K. 2017. "Ancient Maya Beekeeping: The Stingless Bee in Pre-Columbian America." *ThoughtCo*. Available online: https://www.thoughtco.com/ancient-maya-beekeeping-169364 (accessed April 3, 2023).

Hirthammer, M. 2019. "History of Printing Inks." Available online: http://www.blog.wan-ifra.org/articles/2019/04/28/history-of-printing-inks (accessed April 3, 2023).

Hoffmann, R.C. 2005. "A Sixteenth-Century Exemplar and Source." *Environmental History* 10: 702–4.

Hogue, C.L. 1993. *Latin American Insects and Entomology*. Oakland, CA: University of California Press.

Hollingsworth, C. 2001. *Poetics of the Hive: The Insect Metaphor in Literature*. Iowa City: University of Iowa Press.

Hooke, R. 1665. *Micrographia: Or Some Physiological Descriptions of Minute Bodies Made by Magnifying Glasses with Observations and Inquiries thereupon* (fol. a). London: Martyn and Allestry.

Horapollo. 1840. *Hieroglyphics of Horapollo*. Trans. Alexander Turner Cory. London: Chthonios. Available online: http://www.sacred-texts.com/egy/hh/index.htm (accessed April 10, 2023).

Horrox, R., trans. and ed. 1994. *The Black Death*. New York: Manchester University Press.

Houston, K. 2016. *The Book: A Cover-to-Cover Exploration of the Most Powerful Object of Our Time*. New York: W.W. Norton & Company.

Hufthammer, A. and Walløe, L. 2013. "Rats Cannot Have Been Intermediate Hosts for *Yersinia Perstis* During Medieval Plague Epidemics in Northern Europe." *Journal of Archaeological Science* 40: 1751–9.

Hutchinson, G.E. 1974. "Aposematic Insects and the Master of the Brussels Initials." *American Scientist* 62: 161–71.

Ibeji, M. 2011. "Black Death." Available online: http://www.bbc.co.uk/history/british/middle_ages/black_01.shtml (accessed September 8, 2018).

Impelluso, L. 2003. *Nature and its Symbols*. Trans. Stephen Sartarelli. Los Angeles: J. Paul Getty Museum.

James-Chakraborty, K. 2014. *Architecture Since 1400*. Minneapolis: University of Minnesota Press.

Jansen, A.M., Xavier, S.C., and Roque, A.L. 2017. "Ecological Aspects of *Trypanosoma cruzi*: Wild Hosts and Reservoirs." In J. Telleria and A. Tibayrenc (eds.), *American Trypanosomiasis Chagas Disease*, 243–64. London: Elsevier.

Jansen, A.M., Xavier, S.C., and Roque, A.L. 2018. "*Trypanosoma Cruzi* Transmission in the Wild and its Most Important Reservoir Hosts in Brazil." *Parasites & Vectors* 11: 502–27.

Jeffrey, L.N. 1976. "Shelley's 'Plumèd Insects Swift and Free'." *Keats-Shelley Journal* 25: 103–21.

Johns, A. 1998. *The Nature of the Book: Print and Knowledge in the Making*. Chicago: University of Chicago Press.

Johnston, P. 2011. "A Theatre of Insects, or How Nature Lost Her Morality." *AA Files* 63: 36–45.

Jones, D.S. 2003. "Virgin Soils Revisited." *William and Mary Quarterly*, 3rd Series, 60: 703–42.

Jorink, E. 2010. *Reading the Book of Nature in the Dutch Golden Age*. Trans. Peter Mason. Leiden: Brill.

Kaempfer, E. 1999. *Kaempfer's Japan: Tokugawa Culture Observed*. Honolulu: University of Hawaii Press.

Karlsson, G. 1996. "Plague Without Rats: The Case of Fifteenth-Century Iceland." *Journal of Medieval History* 22(3): 263–84.

Kassinger, R.G. 2003. *Dyes: From Sea Snails to Synthetics*. Brookfield, CT: Twenty-First Century Books.

Kaufmann, T.D. 1988. *The School of Prague. Painting at the Court of Rudolf II*. Chicago and London: University of Chicago Press.

Kaufmann, T.D. 1993. *The Mastery of Nature. Aspects of Art, Science, and Humanism in the Renaissance*. Princeton: Princeton University Press.

Kaufmann, T.D. 2009. *Arcimboldo: Visual Jokes, Natural History, and Still-Life Painting*. Chicago and London: University of Chicago Press.

Kellert, S. 1993. "The Biological Basis for Human Values of Nature." In S.R. Kellert and E.O. Wilson (eds.), *The Biophilia Hypothesis*, 42–69. Washington, DC: Island Press.

Kelly, M. and O'Grada, C. 2014. "Living Standards and Mortality in England Since the Middle Ages." *Economic History Review* 32: 358–81.

Kenward, H. and Allison, E. 1994. "Rural Origins of the Urban Insect Fauna." In A.R. Hall and H.K. Kenward (eds.), *Urban-Rural Connexions: Perspectives from Environmental Archaeology*, Oxbow Monograph 47, 55–77. Exeter: Association for Environmental Archaeology.

Kerpel, D.M. 2015. "The Politics of *Nochetzli*: The Florentine Codex and the Nahua World of New Spain." In C. Padilla and B. Anderson (eds.), *A Red Like No Other: How Cochineal Colored the World*, 40–3. New York: Skira Rizzoli.

King, G.A., Kenward, H., Schmidt, E., and Smith, D. 2014. "Six-legged Hitchhikers: An Archaeobiogeographical Account of the Early Dispersal of Grain Beetles." *Journal of the North Atlantic* 23(1): 1–18.

Kirby, J. 2015. "'One of the Most Beautiful Reds': Cochineal in European Painting." In C. Padilla and B. Anderson (eds.), *A Red Like No Other: How Cochineal Colored the World*, 174–89. New York: Skira Rizzoli.

Kircher, A. 1652–1654. *Oedipus Aegypticus*. Rome: V. Mascardi.

Kircher, A. 1675. *Arca Noë*. Amsterdam.

Kleiner, F.S. 2015. *Gardner's Art Through the Ages*, 15th edn. New York: Wadsworth Publishing.

Kline, N.R. 2001. *Maps of Medieval Thought: The Hereford Paradigm*. Woodbridge, Suffolk: The Boyden Press.

Koul, O. and Dhaliwal, G.S., eds. 2001. *Phytochemical Biopesticides*. Amsterdam: Harwood Academic Publishers.

Kramer, H. and Sprenger, J. 1486 (reprinted 1971). *The Malleus Malifacarum of Heinrich Kramer and James Sprenger*. Trans. Montague Summers. New York: Dover.

Kritsky, G. 1997. "The Insects and Other Arthropods of the Bible, the New Revised Version." *American Entomologist* 43(3): 183–8.

Kritsky, G. 2010. *The Quest for the Perfect Hive*. New York: Oxford University Press.

Kritsky, G. 2017. "Beekeeping from Antiquity Through the Middle Ages." *Annual Review of Entomology* 62: 249–64.

Kritsky, G. 2018. "The Quest for the Perfect Hive: Ancient Mediterranean Origins." In F. Hatjina, G. Mavrofridis, and R. Jones (eds.), *Beekeeping in the Mediterranean from Antiquity to the Present*, 50–5. Nea Moudania. Available online: https://issuu.com/fhatjina/docs/beekeeping_in_the_mediterranean_hat (accessed April 3, 2023).

Kritsky, G. and Cherry, R. 2000. *Insect Mythology*. New York: Writers Club Press.

Kritsky, G. and Mader, D. 2010. "Leonardo's Insects." *American Entomologist*, Fall: 178–84.

Kusukawa, S. 2010. "The Sources of Gessner's Pictures for the *Historia animalium*." *Annals of Science* 67(3): 303–28.

Lafita, I., Molina, L., and Molina, C. 2017. "Following the Thread: The Silk Road in Valencia." *Mètode* (Univ. of Valencia) (December 13). Available online: https://metode.org/news/following-the-thread-the-silk-road-in-valencia.html (accessed April 3, 2023).

Lagorio, V.M. 1973. "Excerpts from Pliny's Natural History in Codex Pal. Lat. 1577." *American Journal of Philology* 94: 289–92.

Lambin, E.F., Tran, A., Vanwambeke, S.O., Linard, C., and Soti, V. 2010. "Pathogenic Landscapes: Interactions Between Land, People, Disease Vectors, and Their Animal Hosts." *International Journal of Health Geographics* 9(1): 54.

Lambrecht, F. 1985. "Trypanosomes and Hominid Evolution." *Bioscience* 35: 640–6.
LeCain, R. and Sheley, R. 2014. "Common tansy (*Tanacetum vulgare*)." *Montana State University Extension*. Available online: http://msuinvasiveplants.org/documents/mt_noxious_weeds/common_tansy.pdf (accessed February 28, 2018).
Ledger, M.L. and Mitchell, P.D. 2019. "Tracing Zoonotic Parasite Infections Throughout Human Evolution." *International Journal of Osteoarchaeology* 32(3): 553–64.
Leeson, P.T. 2013. "Vermin Trials." *Journal of Law and Economics* 53(3): 811–36.
Léry, J. de. 1990. *History of a Voyage to the Land of Brazil*. Trans. J. Whatley. Berkeley: University of California Press.
Lesser, F.C. 1738. *Insecto-theology: Or, a Demonstration of the Being and Perfections of God*. Frankfurt and Leipzig: Michael Blochberger.
Leulmi, H., Bitam, I., Berenger, J.M., Lepidi, H., Rolain, J.M., Almeras, L., Raoult, D., and Parola, P. 2015. "Competence of Cimex lectularius Bed Bugs for the Transmission of Bartonella quintana, the Agent of Trench Fever." *PLoS Neglected Tropical Diseases* 9(5): e0003789. DOI: 10.1371/journal.pntd.0003789.
Leverotti, F. 2009. "The Organization of the Sforza Court and Silk Production." In C. Buss (ed.), *Silk Gold Crimson: Secrets and Technology at the Visconti and Sforza Courts*, 18–24. Cinisello Balsamo (Milano): Silvana.
Little, L.K. 2011. "Plague Historians in Lab Coats." *Past and Present* 213(1): 267–90.
López de Gómara, F. [1552] 2003. *Historia General de las Indias*. Available online: http://www.biblioteca.org.ar/libros/92761.pdf (accessed February 13, 2018).
Love, H. 1993. *Scribal Publication in Seventeenth-Century England*. Oxford: Oxford University Press.
Lowood, H. 1995. "The New World and the European Catalog of Nature." In K.O. Kupperman (ed.), *America in European Consciousness, 1493–1750*, 295–323. Chapel Hill: University of North Carolina Press for the Omohundro Institute of Early American History and Culture.
Magnus, O. 1996. *Olaus Magnus: Description of the Northern Peoples, Rome 1555*, vol. 1. Ed. P.G. Foote. London: Hakluyt Society.
Magnus, O. 1998a. *Olaus Magnus: Description of the Northern Peoples, Rome 1555*, vol. 2. Ed. P.G. Foote. London: Hakluyt Society.
Magnus, O. 1998b. *Olaus Magnus: Description of the Northern Peoples, Rome 1555*, vol. 3. Ed. P.G. Foote. London: Hakluyt Society.
Mancall, P.C., ed. 2006. *Travel Narratives from the Age of Discovery*. New York: Oxford University Press.
Mancall, P.C. 2018. *Nature and Culture in the Early Modern Atlantic*. Philadelphia: University of Pennsylvania Press.
Manley, F. 1961. "Milton and the Beasts of the Field." *Modern Language Notes* 76(5): 398–403.
Mann, C.C. 2011. *1493: Uncovering the New World Columbus Created*. New York: Alfred A. Knopf.
Mann, M. 2002. "The Little Ice Age." In M. MacCracken and J. Perry (eds.), *Encyclopedia of Global Environmental Change*. Chichester: John Wiley & Sons, Ltd.
Marciniak, S., Prowse, T.L., Herring, D.A., Klunk, J., Kuch, M. Duggan, Ana T., Bondioli, L., Holmes, E.C., and Poinar, H. 2016. "Plasmodium Falciparum Malaria in 1st–2nd Century CE Southern Italy." *Current Biology* 26(23): PR1220–R1222.

Marichal, C. 2015. "The Cochineal Commodity Chain: Mexican Cochineal and the Rise of Global Trade." In C. Padilla and B. Anderson (eds.), *A Red Like No Other: How Cochineal Colored the World*, 54–67. New York: Skira Rizzoli.
Markley, R. 1993. *Fallen Languages: Crises of Representation in Newtonian England, 1660–1740*. Ithaca: Cornell University Press.
Markley, R. 1999. "Newton, Corruption, and the Tradition of Universal History." In J.E. Force and R. Popkin (eds.), *Newton and Religion*, 121–43. Dordrecht, Netherlands: Kluwer Academic Press.
Marshall, P. 2006. *The Magic Circle of Rudolf II: Alchemy and Astrology in Renaissance Prague*. New York: Walker & Company.
Martinez, A.M. 2016. "Bertilak's Green Vision: Land Stewardship in *Sir Gawain and the Green Knight*." *Arthuriana* 26(4): 114–29.
McDonald, W. 2014. "The World as Woman: Two Late Medieval Song-Poems on Frau Welt by Michel Beheim." *Modern Philology* 111(4): 616–44.
McNeill, J.R. 2010. *Mosquito Empires: Ecology and War in the Greater Caribbean, 1620–1914*. Cambridge: Cambridge University Press.
Megenberg, K. von. 2003. *Das "Buch der Natur" Bd II: Kritischer Text nach den Handschriften*. Ed. R. Luff and G. Steer. Tübingen: Max Niemeyer. [Vol. 2 only, to date.]
Michelet, M. 1882. *History of France*, vol. II. Trans. G.H. Smith. New York: D. Appleton and Company.
Mitchell, P.D. 2015. "Human Parasites in Medieval Europe: Lifestyle, Sanitation and Medical Treatment." *Advances in Parasitology* 90: 389–420.
Mitsuhashi, J. 2016. *Edible Insects of the World*. Boca Raton, FL: CRC Press.
Moffet, T. 1634. *Insectorum sive Minimorum Animalium Theatrum ....* London: Thomas Coates.
Moffet, T. 1658. *The Theater of Insects: Or, Lesser living Creatures, as Bees, Flies, Caterpillars, Spidrs,* [sic] *Worms, &c a most Elaborate Work*. London.
Molà, L. 2003. *The Silk Industry of Renaissance Venice*. Baltimore, MD: Johns Hopkins University Press.
Monnas, L. 2009. *Merchants, Princes, and Painters: Silk Fabrics in Italian and Northern* Paintings, *1300–1550*. New Haven, CT: Yale University Press.
Monnas, L. 2018. "A Study in Contrasts: Silk Consumption in Italy and England in the Fifteenth and Sixteenth Centuries." In D. Schäfer, G. Riello, and L. Móla (eds.), *Threads of Global Desire: Silk in the Pre-Modern World*, 149–86. Rochester, NY: Boydell & Brewer.
Morretto, E. et al. 2010. *Il baco da seta. Storia, cultura, tradizioni e scienza*. Montegrotto Terme (Padua): Butterfly Arc.
Morris, B. 2004. *Insects and Human Life*. Oxford and New York: Berg.
Morris, B. 2008. "Insects as Food Among Hunter-Gatherers." *Anthropology Today* 24(1): 6–8.
Morrison, A.S., Kirshner, J., and Molho, A. 1985. "Epidemics in Renaissance Florence." *American Journal of Public Health* 75(5): 528–35.
Mrak, E. 1969. *Report of the Secretary's Commission on Pesticides and their Relationship to Environmental Health (Parts I and II). U.S.D.H.E.W.* US Govt. Printing Office.
Müster, S. 1546. *Cosmographia*. Basil.
Naruta-Moya, A. 2015. "Interweaving Europe and Central Asia: Connections in Ethnobotany, Folklore, and Polish Cochineal." In C. Padilla and B. Anderson (eds.), *A Red Like No Other: How Cochineal Colored the World*, 202–11. New York: Skira Rizzoli.

Nauert Jr., C.G. 1979. "Humanists, Scientists, and Pliny: Changing Approaches to a Classical Author." *American Historical Review* 84: 72–85.

Neale, J.M. 1856. *Mediaeval preachers and mediaeval preaching: a series of abstracts, translated from the sermons of the middle ages*. London: J. C. Mozley.

Needham, J. 1974. *Science and Civilisation in China*, vol. 5, pt. 2. Cambridge: Cambridge University Press.

Neri, J. 2011. *The Insect and the Image: Visualizing Nature in Early Modern Europe, 1500–1700*. Minneapolis: University of Minnesota Press.

Newfield, T. 2017 "Malaria and Malaria-Like Disease in the Early Middle Ages." *Early Medieval Europe* 25: 251–300.

Newfield, T. 2019. "Mysterious and Mortiferous Clouds: The Climate Cooling and Disease Burden of Late Antiquity." In A. Izdebski and M. Mulryan (eds.), *Environment and Society in the Long Late Antiquity*, 271–97. Leiden: Brill.

Nguyen-Hieu, T., Aboudharam, G., Signoli, M., Rigeade, C., Drancourt, M., and Raoult, D. 2010. "Evidence of a Louse-Borne Outbreak Involving Typhus in Douai, 1710–1712 During the War of Spanish Succession." *PLoS One* 5(10): p.e15405.

Nguyen-Hieu, T., Forestier, C.L., Drancourt, M., Raoult, D., and Aboudharam, G. 2011. "Co-Detection of Bartonella Quintana and Yersinia Pestis in an 11th–15th Burial Site in Bondy, France." *American Journal of Physical Anthropology* 145(3): 489–94.

Niu, J.-Z., Hull-Sanders, H., Zhang, Y.-X., Lin, J.-Z., Dou, W., and Wang, J.-J. 2014. "Biological Control of Arthropod Pests in Citrus Orchards in China." *Biological Control* 68: 15–22.

Nohl, J. 2006. *The Black Death*. Trans. C.H. Clarke. Yardley, PA: Westholme Publishing, LLC.

Novo, S., Leles, D., Bianucci, R., and Araujo, A. 2016. "The Process of Leishmania Infection – Disease and New Perspectives of Paleoparasitology." *Revista do Instituto de Medicina Tropical de São Paulo* 58: 45. Available online: https://www.ncbi.nlm.nih.gov/pmc/articles/PMC4880002/ (accessed April 3, 2023).

Nutton, V. 1983. "The Seeds of Disease: An Explanation of Contagion and Infection from the Greeks to the Renaissance." *Medical History* 27: 1–34.

Nutton, V. 2008. "Pestilential Complexities: Understanding Medieval Plague." *Medical History*, Supplement 27: 1–16.

Ogilvie, B.W. 2006. *The Science of Describing: Natural History in Renaissance Europe*. Chicago: University of Chicago Press.

Ogilvie, B.W. 2008. "Nature's Bible: Insects in Seventeenth-Century Art and Science." *Tidsskrift fur kulturforskning* 16(3): 5–21.

Ogilvie, B.W. 2012. "Attending to Insects: Francis Willughby and John Ray." *Notes and Records of the Royal Society of London* 66: 357–72.

Olbricht, E.M. 2006. "Made Without Hands: The Representation of Labor in Early Modern Silkworm and Beekeeping Manuals." In E.C. Brown (ed.), *Insect Poetics*, 223–41. Minneapolis and London: University of Minnesota Press.

Oonincx, D.G.A.B. 2015. "Insects as Food and Feed: Nutrient Composition and Environmental Impact." PhD thesis, Wageningen University, Wageningen, NL.

Oosthuizen, M.J. 1935. *The Effect of High Temperature on the Confused Flour Beetle*. University of Minnesota Agricultural Experiment Station, Technical Bulletin. Available online: https://books.google.com/books/about/The_Effect_of_High_Temperature_on_the_Co.html?id=F4xAAAAAIAAJ (accessed February 3, 2017).

Otto, R. [1923] 1958. *The Idea of the Holy*. Oxford: Oxford University Press.

Pächt, O. 1950. "Early Italian Nature Studies and the Early Calendar Landscape." *Journal of the Warburg and Courtauld Institutes* 13(1/2): 13–47.
Padilla, C. and Anderson, B., eds. 2015. *A Red Like No Other: How Cochineal Colored the World*. New York: Skira Rizzoli.
Pagden, A. 1986. *The Fall of Natural Man: The American Indian and the Origins of Comparative Ethnology*. Cambridge: Cambridge University Press.
Panagiotakopulu, E. and Buckland, P.C. 2017. "A Thousand Bites—Insect Introductions and Late Holocene Environments." *Quaternary Science Reviews* 156: 23–35.
Panagiotakopulu, E., Buckland, P.C., and Wickler, S. 2018. "Is There Anybody in There? Entomological Evidence from a Boat Burial at Øksnes in Vesterålen, Northern Norway." *PloS One* 13(7): p.e0200545.
Paré, A. 1982. *Ambroise Paré on Monsters and Marvels*. Trans. Janis L. Pallister. Chicago: University of Chicago Press.
Parikka, J. 2010. *Insect Media: An Archaeology of Animals and Technology*. Minneapolis: University of Minnesota Press.
Parsons, J.R. 2005. "The Aquatic Component of Aztec Subsistence: Hunters, Fishers, and Collectors in an Urbanized Society." *Michigan Discussions in Anthropology* 15(1). Available online: http://hdl.handle.net/2027/spo.0522508.0015.104 (accessed February 13, 2018).
Parsons, J. 2010. "The Pastoral Niche in Pre-Hispanic Mesoamerica." In J.E. Staller and M.D. Carrasco (eds.), *Pre-Columbian Foodways: Interdisciplinary Approaches to Food, Culture, and Markets in Mesoamerica*, 109–36. New York: Springer.
Patterson, R. 1838. *Letters on the Natural History of the Insects Mentioned in Shakespeare's Plays, with Incidental Notices of the Entomology of Ireland*. London: Wm. S. Orr and Co., Paternoster Row. Available online: https://www.biodiversitylibrary.org/bibliography/33441#/summary (accessed April 3, 2023).
Pauketat, T.R. 2009. *Cahokia: Ancient America's Great City on the Mississippi*. New York: Viking.
Payne, J. [1886] 2017. *The Collected Works of Giovanni Boccacio,* vol. 2: *The Decameron John Payne 1886*. Trans. John Payne. Hastings, UK: Delphi Classics.
Peigler, R.S. 1993. "Wild Silks of the World." *American Entomologist* 39(3): 151–61.
Peverly, S. 2014. "Iron Gall Ink: A Medieval Recipe." Available online: https://sarahpeverley.com/2014/01/29/iron-gall-ink-a-medieval-recipe/ (accessed July 2, 2019).
Phillips, A. 2018. "The Localization of the Global: Ottoman Silk Textiles and Markets, 1500–1790." In D. Schäfer, G. Riello, and L. Móla (eds.), *Threads of Global Desire: Silk in the Pre-Modern World*, 103–23. Rochester, NY: Boydell & Brewer.
Phipps, E. [2010] 2012. *Cochineal Red: The Art History of a Color*. New Haven and London: The Metropolitan Museum of Art, Yale University Press.
Phipps, E. 2015. "Shades of Red: Color and Culture in Andean Textiles." In C. Padilla and B. Anderson (eds.), *A Red Like No Other: How Cochineal Colored the World*, 106–17. New York: Skira Rizzoli.
Pliny the Elder. 1855. *The Natural History*, vol. 3, book 11, Ch. 4. Trans. John Bostock. London: Taylor and Francis. Available online: http://www.perseus.tufts.edu/hopper/text?doc=Perseus%3Atext%3A1999.02.0137%3Abook%3D11%3Achapter%3D4 (accessed June 3, 2018).
Pliny the Elder. 1938. Natural History, vol. 3, book 11, Ch. 1. Trans. H. Rackham, W. Jones, and D. Eichholz. London: William Heineman. Available online: https://en.wikisource.org/wiki/Natural_History_(Rackham,_Jones,_%26_Eichholz)/Book_11 (accessed April 3, 2023).

Ponting, C. 2011. *A New Green History of the World: The Environment and the Collapse of Great Civilizations*. New York: Random House.
Poos, L.R. 1989. "The Historical Demography of Renaissance Europe: Recent Research and Current Issues." *Renaissance Quarterly* 42: 794–811.
Potter, P. 2013. "The Iconography of Vermin." *Emerging Infectious Diseases* 19(2): 350–1.
Powell, J.R. and Tabachnick, W.J. 2013. "History of Domestication and Spread of *Aedes aegypti*—a Review." *Memórias do Instituto Oswaldo Cruz* 108: 11–17.
Purchas, S. 1657. *A theater of politicall flying-insects*. London: R.I. for Thomas Parkhurst.
Pyle, C.M. 1984. *Das Tierbuch des Petrus Candidus. Eine Einführung*. Zürich: Belser Verlag (Codices e Vaticanis Selecti, LX).
Pyle, C.M. 1996a. "The Art and Science of Renaissance Natural History: Thomas of Cantimpré, Pier Candido Decembrio, Conrad Gessner and Teodoro Ghisi in Vatican Library MS Urb. lat. 276." *Viator* 27: 265–321.
Pyle, C.M. 1996b. "Teodoro Ghisi." In J.S. Turner (ed.), *The Dictionary of Art*, 12: 558. London, Macmillan. Revised in *The Encyclopedia of Italian Renaissance and Mannerist Art*, 1: 645–6. London: Macmillan.
Pyle, C.M. 2000. "Art as Science: Scientific Illustration, 1490–1670, in Drawing, Woodcut and Copper Plate." *Endeavour* 24: 69–75.
Pyle, C.M. 2001. "Art as Science in Joris Hoefnagel and Teodoro Ghisi." *Center 21. Record of Activities and Research Reports, June 2000–May 2001*. Washington: Center for Advanced Study in the Visual Arts, National Gallery of Art: 112–14.
Pyle, C.M. 2005. "Thomas of Cantimpré." In T. Glick, S. J. Livesey, and F. Wallis (eds.), *Medieval Science, Technology and Medicine: An Encyclopedia*, 476–8. New York and London: Routledge.
Pyle, C.M. 2006. "The Renaissance Rediscovery of the Classical Approaches to the World: Reflections on History and Science, Then and Now." In R. Suntrup and J.R. Veenstra (eds.), *Building the Past/Konstruktion der eigenen* Vergangenheit, 3–31. Frankfurt am Main, Berlin, New York, etc.: Peter Lang.
Quarles, F. [1635] 1643. *Emblemes by Francis Quarles*. Cambridge: Printed by R.D. for Francis Eglesfield.
Radford, U.M. 1949. "The Wax Images Found in Exeter Cathedral." *Antiquaries Journal* 29(3–4): 164–8.
Raffles, H. 2010. *Insectopedia*. New York: Pantheon Books.
Ramírez, J.A. 2000. *The Beehive Metaphor: From Gaudí to Le Corbusier*. Trans. A.R. Tulloch. London: Reaktion Books.
Ramón Y Cajal, S. 2017. *The Beautiful Brain. The Drawings of Santiago Ramón y Cajal*. Ed. E.A. Newman, A. Araque, and J.M. Dubinsky. New York: Abrams.
Ramos-Elorduy, J., Pino Moreno, J.M., and Mayaudon, C.M. 1984. "Protein Content of Some Edible Insects in Mexico." *Journal of Ethnobiology* 4: 61–72.
Ramos-Elorduy, J. and Pino Moreno, J.M. 1989a. *Los Insectos comestibles en el México antiguo: estudio etnoentomológico*. AGT. Available online: https://books.google.com/books/about/Los_Insectos_comestibles_en_el_México_a.html?id=Hx4wAAAACAAJ (accessed February 13, 2018).
Ramos-Elorduy, J. and Pino Moreno, J.M. 1989b. "Honduras is Great. ¿Sabía que los zompopos de mayo se comen?" Available online: https://hondurasisgreat.org/zompopos-honduras/ (accessed August 14, 2019).
Ramos-Elorduy, J., Luis, J., and Montesinos, V. 2007. "Los insectos como alimento humano: Breve ensayo sobre la entomofagia, con especial referencia a México"

[Insects as human food: short essay on entomophagy, with special reference to Mexico.] *Boletín de la Real Sociedad Española de Historia Natural*. Sección Biológica 102(4): 61–84.

Ransome, H. 2004. *The Sacred Bee in Ancient Times and Folklore*. Mineola, NY: Dover Publications, Inc.

Raoult, D. 2016. "A Personal View of How Palaeomicrobiology Aids Our Understanding of the Role of Lice in Plague Pandemics." *Microbiology Spectrum* 4.

Raoult, D., Aboudharam, G., Crubezy, E., Larrouy, G., Ludes, B., and Drancourt, M. 2000. "Molecular Identification by 'Suicide PCR' of *Yersinia pestis* as the Agent of Medieval Black Death." *Proceedings of the National Academy of Sciences* 97(23): 12800–3.

Raoult, D., Dutour, O., Houhamdi, L., Jankauskas, R., Fournier, P.E., Ardagna, Y., Drancourt, M., Signoli, M., La, V.D., Macia, Y., and Aboudharam, G. 2006. "Evidence for Louse-Transmitted Diseases in Soldiers of Napoleon's Grand Army in Vilnius." *The Journal of Infectious Diseases* 193(1): 112–20.

Raoult, D., Woodward, T., and Dumler, J.S. 2004. "The History of Epidemic Typhus." *Infectious Disease Clinics of North America* 18(1): 127–40.

Ratcliffe, B.C. 2006. "Scarab Beetles in Human Culture." *Coleopterists Society Monograph* 5: 85–101. Available online: http://digitalcommons.unl.edu/entomologypapers/94/ (accessed April 3, 2023).

Raven, C.E. 1947. *English Naturalists from Neckham to Ray: A Study of the Making of the Modern World*. Cambridge: Cambridge University Press.

Rawcliffe, C. 2012. "Sources for the Study of Public Health in the Medieval City." In J. Rosenthal (ed.), *Understanding Medieval Primary Sources: Using Historical Sources to Discover Medieval Europe*, 177–95. London: Routledge Publishing.

Rawcliffe, C. 2013. *Urban Bodies: Public Health in Late Medieval English Towns and Cities*. Woodbridge, UK: Boydell Press.

Ray, J. 1691. *The wisdom of God manifested in the works of the creation*. London: Samuel Smith.

Reisen, W.K. 2010. "Landscape Epidemiology of Vector-Borne Diseases." *Annual Review of Entomology* 55: 461–83.

Reiter, P. 2000. "From Shakespeare to Defoe: Malaria in England in the Little Ice Age." *Emerging Infectious Diseases* 6(1): 1–11.

Reiter, P. 2001. "Climate Change and Mosquito-Borne Disease." *Environmental Health Perspectives* 109 (supplement 1): 141–61.

Reyerson, K.L. 1982. "Medieval Silks in Montpellier: The Silk Market ca. 1250–1350." *Journal of European Economic History* 11: 117–40.

Reyerson, K.L. 2013. "Urban Economies." In J.M. Bennett and R.M. Karras (eds.), *The Oxford Handbook of Women and Gender in Medieval Europe*, 295–310. London: Oxford University Press.

Riley, J.C. 1986. "Insects and the European Mortality Decline." *American Historical Review* 91: 833–58.

Ripa, C. [1603] 1709. *Iconologia, or, Moral emblems*. London: Printed by Benj. Motte.

Roberts, L., Janovy, Jr., J., and Nadler, S. 2013. *Foundations of Parasitology*, 9th edn. New York: McGraw-Hill Publishers.

Robertshaw, P. 1998. "Medieval Household Pest Control." Available online: http://www.granta.demon.co.uk/arsm/jg/pest.html (accessed January 6, 2018; URL no longer active).

Ross, A. 1626. *An Exposition on the Fourteene First Chapters of Genesis*. London: Printed by B(ernard) A(lsop) and T(homas) F(awcett).
Routt, D. 2017. "The Economic Impact of the Black Death." In R.D. Johnson (ed.), *An Introductory To Economics: Key Concept Summaries and Topics in Microeconomics and Macroeconomics*, 41–59. Lulu.com.
Rowland, I. 2002. "Athanasius Kircher and His Egyptian Oedipus." *Fathom Archive*. Available online: http://fathom.lib.uchicago.edu/1/777777122590/ (accessed June 1, 2018).
Rowland, I. 2005. "Poetry and Prophecy in the Encyclopedic System of Athanasius Kircher." *Bruniana & Campanelliana* 11(2): 509–17.
Sadler, J.P. 1990. "Records of Ectoparasites on Humans and Sheep from Viking-Age Deposits in the Former Western Settlement of Greenland." *Journal of Medical Entomology* 27: 628–31.
Sahagún, B. [1540–1585] 1950–1982. *General History of the Things of New Spain: Florentine Codex*. Trans. and ed. A.J.O. Anderson and C.E. Dibble. 13 parts. Santa Fe and Salt Lake City: School of American Research and University of Utah Press.
Salviani, I. 1554. *Aquatilium Animalium Historiae liber primvs, cvm eorvmedem formis, ære exevsis*. Romæ: Hippolyto Salviano typhernate Romæ medicinam profitente auctore.
Saunders, M. 2016. "Shakespeare Knew His Insects." Available online: https://ecologyisnotadirtyword.com/2016/04/26/shakespeare-insects/ (accessed Fenbruary 26, 2020).
Scarborough, J. 1977. "Some Beetles in Pliny's Natural History." *The Coleopterists Bulletin* 31: 293–6.
Schats, R. 2015. "Malaise and Mosquitos: Osteoarchaeological Evidence for Malaria in the Medieval Netherlands." *Analecta Praehistorica Leidensia* 45: 133–40.
Schelvis, J. 1994. "Caught Between the Teeth. A Review of Dutch Finds of Archaeological Remains of Ectoparasites in Combs." In M.J. Sommeijer and J. van der Blom (eds.), *Experimental and Applied Entomology (Proceedings of the Netherlands Entomological Society)* 5: 131–2.
Schedel, H. 2010. *Liber chronicarum ….* (1493) *First English Edition of the Nuremberg Chronicle*. Madison: University of Wisconsin. Available online: http://digicoll.library.wisc.edu/cgi/t/text/text-idx?c=nur;cc=nur;rgn=div2;view=text;idno=nur.001.0004;node=nur.001.0004%3A8.285 (accessed April 3, 2023).
Schmetzer, U. 1988. "'Worst Locust Plague in History' Threatens Europe." *Chicago Tribune*, May 2. Available online: http://articles.chicagotribune.com/1988-05-02/news/8803130590_1_swarms-desert-locust-locust-larvae (accessed March 2, 2017).
Schmid, B.V., Büntgen, U., Easterday, W.R., Ginzler, C., Walløe, L., Bramanti, B., and Stenseth, N.C. 2015. "Climate-Driven Introduction of the Black Death and Successive Plague Reintroductions into Europe." *Proceedings of the National Academy of Sciences of the United States of America* (March 10) 112(10): 3020–5.
Sealey, N. 1991. "An Examination of the Geography of Three Major Contenders for Columbus's First Landfall, in 1492." *Terrae Incognitae* 12: 39–50.
Seifert, L., Harbeck, M., Thomas, A., Hoke, N., Zöller, L., Wiechmann, I., Grupe, G., Scholz, H.C., and Riehm, J.M. 2013. "Strategy for Sensitive and Specific Detection of *Yersinia Pestis* in Skeletons of the Black Death Pandemic." *PLoS One* 8(9): e75742. DOI: 10.1371/journal.pone.0075742.
Seifert, L., Wiechmann, I., Harbeck, M., Thomas, A., Grupe, G., Projahn, M., Scholz, H.C., and Riehm, J.M. 2016. "Genotyping *Yersinia Pestis* in Historical

Plague: Evidence for Long-Term Persistence of *Y. Pestis* in Europe from the 14th to the 17th Century." *PLoS One* January 13, 11(1): e0145194. DOI: 10.1371/journal.pone.0145194. PMID: 26760973; PMCID: PMC4712009.

Sereno, D., Akhoundi, M., Dorkeld, F., Oury, B., Momen, H., and Perrin, P. 2017. "What Pre-Columbian Mummies Could Teach Us About South American Leishmaniases?" *Pathogens and Disease* 75(3) (April): ftx019. Available online: https://doi.org/10.1093/femspd/ftx019 (accessed April 3, 2023).

Sethi, C.M. 2015. "To Dye For: The Many Reds of Asia." In C. Padilla and B. Anderson (eds.), *A Red Like No Other: How Cochineal Colored the World*, 68–71. New York: Skira Rizzoli.

Shakespeare, W. [1595] 2016. *A Midsummer Night's Dream*. In S. Greenblatt et al. (eds.), *The Norton Shakespeare*. New York/London: W.W. Norton & Company.

Shakespeare, W. [1595] 2016. *Richard II*. In S. Greenblatt et al. (eds.), *The Norton Shakespeare*. New York/London: W.W. Norton & Company.

Shakespeare, W. [1594] 2016. *Romeo and Juliet*. In S. Greenblatt et al. (eds.), *The Norton Shakespeare*. New York/London: W.W. Norton & Company.

Shakespeare, W. [1591–1594] 2002. *Titus Andronicus*. In S. Orgel and A.R. Braunmuller (eds.), *The Complete Pelican Shakespeare*. New York: Penguin Books.

Shakespeare, W. [1589–1591] 2016. *Two Gentlemen of Verona*. In S. Greenblatt et al. (eds.), *The Norton Shakespeare*. New York/London: W.W. Norton & Company.

Shelley, M. 2020. "Joris Hoefnagel's Insects." Metropolitan Museum Journal 55: 26–42.

Si, A. and Turpin, M. 2015. "The Importance of Insects in Australian Aboriginal Society: A Dictionary Survey." *Ethnobiology Letters* 6: 175–82.

Simond, M., Godley, M., and Mouriquand, P. 1998. "Paul-Louis Simond and His Discovery of Plague Transmission by Rat Fleas: A Centenary." *Journal of the Royal Society of Medicine* 91: 101–4.

Sinden, R.E. 1999. "*Plasmodium* Differentiation in the Mosquito." *Parassitologia* 41(1–3): 139–48.

Sinka, M.E., Bangs, M.J., Manguin, S., Rubio-Palis, Y., Chareonviriyaphap, T., Coetzee, M., Mbogo, C.M., Hemingway, J., Patil, A.P., Temperley, W.H., Gething, P.W., Kabaria, C.W., Burkot, T.R., Harbach, R.E., and Hay, S.I. 2012. "A Global Map of Dominant Malaria Vectors." *Parasit Vectors* 5: 69.

Slavin, P. 2019. "Death by the Lake: Mortality Crisis in Early Fourteenth-Century Central Asia." *Journal of Interdisciplinary History* 50(1): 59–90.

Sleigh, C. 2006. "Inside Out: The Unsettling Nature of Insects." In E.C. Brown (ed.), *Insect Poetics*, 281–97. Minneapolis and London: University of Minnesota Press.

Smith, A.F. 2013. *The Oxford Encyclopedia of Food and Drink in America*. Oxford: Oxford University Press.

Smith, J.C. 1983. *Nuremberg. A Renaissance City 1500–1618*. Austin: University of Texas Press.

Smith, L. 2019. "Early Modern Nitpicking." The Recipe's Project. Available online: https://recipes.hypotheses.org/14901 (accessed April 29, 2019).

Smith, M.E. 2005. "City Size in Late Postclassic Mesomerica." *Journal of Urban History* 31: 403–34.

Smith, P.H. 2006. *The Body of the Artisan: Art and Experience in the Scientific Revolution*. Chicago and London: The University of Chicago Press.

Smith, P.J. 2017. "Joachim Camerarius's Emblem Book on Birds (1596), with an Excursus on America's Great Seal." In K.A.E. Enenkel and P.J. Smith (eds.), *Emblems and the Natural World*, 149–83. Leiden: Brill.

Smithsonian Tropical Research Institute. 2005. "Mayan Stingless Bee Keeping: Going, Going, Gone?" *ScienceDaily*, June 16. Available online: www.sciencedaily.com/releases/2005/06/050615062105.htm (accessed April 10, 2018).

Socolovschi, C. and Raoult, D. 2009. "Typhus Fevers and Other Rickettsial Diseases, Historical." In M. Schaechter (ed.), *Encyclopedia of Microbiology*, 3rd edn. Amsterdam: Elsevier/Academic Press.

Sorré, L. n.d. "Christopher Columbus ... His Gastronomic Persona." Available online: http://questgarden.com/104/72/6/100609065803/files/ChristopherColumbus-HisGastronomicPersona.pdf (accessed February 22, 2018; URL no longer active).

Southerne, E. 1593. *A Treatise Concerning the Right Use and Ordering of Bees: Newlie Made and Set Forth, According to the Author's Owne Experience.: (which by any heretofore hath not been done)*. London: Thomas Owen for Thomas Woodocke. Available online: https://babel.hathitrust.org/cgi/pt?id=wu.89098857493&view=1up&seq=1 (accessed April 3, 2023).

Spindle, E. 2018. "The Story of Cinnabar and Vermilion (HgS) at The Met." The Metropolitan Museum of Art [blog]. Available online: https://www.metmuseum.org/blogs/collection-insights/2018/cinnabar-vermilion (accessed August 17, 2019).

Spyrou, M.A., Keller, M., Tukhbatova, R.I., Scheib, C.L., Nelson, E.A., Valtueña, A.A., Neumann, G.U., Walker, D., Alterauge, A., Carty, N., and Cessford, C. 2019. "Phylogeography of the Second Plague Pandemic Revealed Through Analysis of Historical *Yersinia pestis* Genomes." *Nature Communications* 10(1): 1–13.

St. Anthony of Padua. 1867. *The Moral Concordances of St. Anthony of Padua*. Trans. Rev. J.M. Neale. London: J.T. Hayes.

Starna, W.A., Hamell, G.R., and Butts, W.L. 1984. "Northern Iroquoian Horticulture and Insect Infestation: A Cause for Village Removal." *Ethnohistory* 31(3): 197–207.

Steele, R. 1803. *Mediaeval Lore from Bartholomew Anglicus*. London: Alexander Moring.

Stenseth, N.C., Samia, N.I., Viljugrein, H., Kausrud, K.L., Begon, M., Davis, S., Leirs, H., Dubyanskiy, V.M., Esper, J., Ageyev, V.S., Klassovskiy, N.L., Pole, S.B., and Chan, K. 2006. "Plague Dynamics are Driven by Climate Variation." *PNAS* 103(35): 13110–15.

Steverding, D. 2008. "The History of African Trypanosomiases." *Parasit Vectors* 1: 3. Available online: https://www.ncbi.nlm.nih.gov/pmc/articles/PMC2270819/ (accessed April 3, 2023).

Steverding, D. 2014. "The History of Chagas Disease." *Parasit Vectors* 7: 317. Available online: https://doi.org/10.1186/1756-3305-7-317 (accessed April 3, 2023).

Stige, L.C., Chan, K.S., Zhang, Z., Frank, D., and Stenseth, N.C. 2007. "Thousand-Year-Long Chinese Time Series Reveals Climatic Forcing of Decadal Locust Dynamics." *Proceedings of the National Academy of Sciences USA* 104(41): 16188–93.

Stolzenberg, D. 2013. *Egyptian Oedipus: Athanasius Kircher and the Secrets of Antiquity*. Chicago: University of Chicago Press.

Sussman, G.D. 2011. "Was the Black Death in India and China?" *Bulletin of the History of Medicine* 85(3): 319–55.

Swinburne, L. 1997. "Dancing with the Mermaids: Ship's Biscuit and Portable Soup." In H. Walker (ed.), *Food on the Move: Proceedings of the Oxford Symposium on Food and Cookery, 1996*, 309–20. London: Prospect Books.

Teixeira, A.R., Nitz, N., Guimaro, M.C., Gomes, C., and Santos-Buch, C.A. 2006. "Chagas Disease." *Postgraduate Medical Journal* 82(974): 788–98.

Tellini Perina, C. 1979. "Teodoro Ghisi: L'immagine fra Maniera e Controriforma." In D.A. Franchini, R. Margonari, G.G. Olmi, R. Signorini, A. Zanca, and C.T. Perina, *La Scienza a Corte. Collezionismo eclettico, natura e immagine a Mantova fra Rinascimento e Manierismo*, 239–68. Rome: Bulzoni.

Thacker, J. 2002. *An Introduction to Arthropod Pest Control*. Cambridge: Cambridge University Press.

The Silk Industry of Lyon. n.d. Available online: https://belovedlinens.net/fabrics/m.Lyon-silk1.html (accessed June 10, 2019).

The Statute of Laborers. 1351. In E.F. Henderson, *Select Historical Documents of the Middle Ages*, 307. London: George Bell and Sons, 1896. Available online: http://avalon.law.yale.edu/medieval/statlab.asp (accessed February 22, 2018).

Thirsk, J. 1997. *Alternative Agriculture: A History from the Black Death to the Present Day*. Oxford: Oxford University Press.

Thompson, J.E.S. 1990. *Maya History and Religion*, vol. 99. Norman, OK: University of Oklahoma Press.

Thorndike, L. 1936. "Epitomes of Pliny's Natural History in the Fifteenth Century." *Isis* 26: 39.

Tian, H., Stige, L.C., Cazelles, B., Kausrud, K.L., Svarverud, R., Stenseth, N.C., and Zhang, Z. 2011. "Reconstruction of a 1,910-Y-Long Locust Series Reveals Consistent Associations With Climate Fluctuations in China." *Proceedings of the National Academy of Sciences USA* 108(35): 14521–6.

Tognetti, S. 2005. "The Development of the Florentine Silk Industry: A Positive Response to the Crisis of the Fourteenth Century." *Journal of Medieval History* 31: 55–69.

Tomlinson, R.C. 2012. "'Plusieurs chose qu'il n'avoit veuës': Antoine Du Pinet's Translation of Pliny the Elder (1562)." *Translation and Literature* 21: 145–61.

Topsell, E. 1607. *The Historie of Foure-Footed Beastes*. London.

Topsell, E. 1658. *The history of four-footed beasts and serpents ... whereunto is now added, The theater of insects, or, Lesser living creatures ... by T. Muffet*. London: E. Cotes.

Topsell, E. 1967. *The History of Four-Footed Beasts and Serpents and Insects* (1658). 3 vols. Introduction by Willy Ley. New York: Da Capo.

Tung, M. (1621). *Andrea Alciato's Emblemata*. Padua: Petro Paulo Tozzi. Available online: http://www.emblems.arts.gla.ac.uk/alciato/picturae.php?id=A21a149 (accessed June 16, 2018).

Tusser, T. 1580. *A Hundreth Points of Good Husbandrie*. London: R. Tottel.

Ultee, M. 1988. "Sir Hans Sloane, Scientist." *The British Library Journal* (Sloane Issue) 14: 1–20.

USS Constitution Museum. 2014. "Some Notes on Navy Biscuit." Available online: https://ussconstitutionmuseum.org/2014/07/16/some-notes-on-navy-biscui/ (accessed April 3, 2023).

Van Dam, A., Martínez, L., Chávez, A., and May, B. 2015. "Point of Origin: Genetic Diversity and the Biogeography of the Cochineal Insect." In C. Padilla and B. Anderson (eds.), *A Red Like No Other: How Cochineal Colored the World*, 90–3. New York: Skira Rizzoli.

Van Steesel, A. 2016. "Competing Cities and Urban Networks in Medieval Europa." In V. Mamadouh and A. van Wageningen (eds.), *Urban Europe*, 111–16. Amsterdam: Amsterdam University Press.

Varlik, N. 2015. *Plague and Empire in the Early Modern Mediterranean World: The Ottoman Experience, 1347–1600*. New York: Cambridge University Press.

Veale, E.M. 1957. "The Rabbit in England." *The Agricultural History Review* 5(2): 85–90.

Vicentini, C.B., Manfredini, S., and Contini, C. 2018. "Ancient Treatment for Lice: A Source of Suggestions for Carriers of Other Infectious Diseases?" *Le Infezioni in Medicina* 26(2): 181–92. PMID: 29932096.

Vignau-Wilberg, T. 1994. *Archetypa Studiaque Patris Georgii Hoefnagelii 1592. Natur, Dichtung und Wissenschaft in der Kunst um 1600. Nature, Poetry and Science in Art around 1600*. Munich: Staatliche Graphische Sammlung.

Vinci, L. da. 1898. *I manoscritti di Leonardo da Vinci della Reale Biblioteca di Windsor: Dell'Anatomia – Fogli A*. Ed. T. Sabachnikoff and G. Piumati. Turin: Roux e Viarengo.

Virgil. 1507. *Georgics, Aeneis*. Available online: http://daten.digitale-sammlungen.de/0001/bsb00010391/images/index.html?fip=193.174.98.30&seite=313&pdfseitex= (accessed April 3, 2023).

Von Engeln, O.D. 1916. "Shakespeare, the Observer of Nature." *The Scientific Monthly* 2(6): 573–88.

Von Megenberg, K. 1481. *Das Buch der Natur*. Augsburg: Johann Bämler. Available online: https://www.loc.gov/resource/rbc0001.2008rosen0080/?sp=1&st=list (accessed April 3, 2023).

Wade, M. n.d. "What Is an Emblem?" *Emblematica Online*. Available online: http://emblematica.grainger.illinois.edu (accessed June 12, 2018).

Warner, F. 1921. *The Silk Industry of the United Kingdom: Its Origin and Development*. London: Drane's.

Weisner, M.E. 1986. "Spinsters and Seamstresses: Women in Cloth and Clothing Production." In M.W. Ferguson, M. Quilligan, and N. Vickers (eds.), *Rewriting the Renaissance: The Discourses of Sexual Difference in Early Modern Europe*, 191–205. Chicago: University of Chicago Press.

Weiss, B.B. 1927. "Encyclopedic Entomologists of the Renaissance." *Journal of the New York Entomological Society* 35: 193–9.

Weiss, H. 1930. "Insects and Witchcraft." *Journal of the New York Entomological Society* 38(2): 127–33.

Wheeler, J. 2009. *Renaissance Secrets: Recipes & Formvlas*. London: V&A Publishing.

Wheelis, M. 2002. "Biological Warfare at the Siege of Caffa." *Emerging Infectious Diseases* 8(9): 971–5.

White, A.J., Stevens, Lora R., Lorenzi, V., Munoz, Samuel E., Schroeder, S., Cao, A., and Bogdanovich, T. 2019. "Fecal Stanols Show Simultaneous Flooding and Seasonal Precipitation Change Correlate With Cahokia's Population Decline." *PNAS* 116: 5461–6.

Whitney, G. 1586. *Choice of Emblemes*. Leiden: Christopher Plantin. Available online: http://www.mun.ca/alciato/whit/w200.html (accessed June 16, 2018).

Wilkins, J. 1668. *An Essay Towards a Real Character, And a Philosophical Language*. London: Printed for S.A. Gellibrand and for John Martyn.

Willis, T. 1683. *Two Discourses Concerning the Soul of Brutes*. London: Thomas Dringe.

Wilson, E.O. 2005. "Environment: Early Ant Plagues in the New World." *Nature* 433: 32.

Wilson, T. 1612. *A Christian dictionarie Opening the signification of the chiefe words dispersed generally through the Old and New Testament*. London: W[illiam] Iaggard.

Wither, G. 1628. *Britain's Remembrancer* (1628). Reprint: Manchester: Charles Simms, for the Spenser Society, 1880.
Wolfe, H. "Early Modern Head Lice Remedies." The Collation, May 15. Available online: https://collation.folger.edu/2018/05/early-modern-head-lice-remedies/ (accessed April 33, 2023).
Woolfson, J. 2010. "The Renaissance of Bees." *Renaissance Studies* 24: 281–300.
World Health Organization. 2017. "Plague." Available online: https://www.who.int/news-room/fact-sheets/detail/plague (accessed December 12, 2019).
World Health Organization. 2020. "Trypanosomiasis, Human African (Sleeping Sickness)." Available online: https://www.who.int/news-room/fact-sheets/detail/trypanosomiasis-human-african- (sleeping-sickness) (accessed January 7, 2020).
Yale Library. 2013. *Iron Gall Ink*. Available online: https://travelingscriptorium.library.yale.edu/2013/03/21/iron-gall-ink/ (accessed April 3, 2023).
Yeola, L.H., Maru, A.D., Surwase, R.K., and Shinkar, D.M. 2018. "A Review on Lipstick by Using Indigenous Vegetable Wax." *World Journal of Pharmacy and Pharmaceutical Sciences* 7(6): 1437–46.
Zanoboni, M.P. 2009. "Female Labour in the Silk Industry." In C. Buss (ed.), *Silk Gold Crimson: Secrets and Technology at the Visconti and Sforza Courts*, 33–6. Cinisello Balsamo (Milano): Silvana.
Zeigler, P. 1969. *The Black Death*. New York: HarperCollins Publishers.
Ziegler, M. 2016a. "The Black Death and the Future of the Plague." *The Medieval Globe* 1(1): 259–83.
Ziegler, M. 2016b. "The Promiscuous Human Flea." Contagions. Available online: https://contagions.wordpress.com/2016/09/27/the-promiscuous-human-flea/ (accessed April 3, 2023).
Ziegler, M. 2016c. "Landscapes of Disease." *Landscapes* 17(2): 99–107.
Zink, A.R., Spiegelman, M., Schraut, B., Greenblatt, C.L., Nerlich, A.G., and Donoghue, H.D. 2006. "Leishmaniasis in Ancient Egypt and Upper Nubia." *Emerging Infectious Diseases* 12(10): 1616–17.

INDEX